the butterfly handbook

the butterfly handbook

Dr Jacqueline Y. Miller • Dr Lee D. Miller

BARRON'S

A QUARTO BOOK

First edition for the United States, its territories and dependencies, and Canada published in 2004 by Barron's Educational Series, Inc.

All inquiries should be addressed to:
Barron's Educational Series, Inc.
250 Wireless Boulevard, Hauppauge,
New York 11788
http://www.barronseduc.com

International Standard Book No.
0-7641-5714-0

Library of Congress Control No: 2003107440

QUAR.BUH

Conceived, designed, and produced by
Quarto Publishing plc
The Old Brewery
6 Blundell Street
London N7 9BH

PROJECT EDITOR: Jocelyn Guttery
ART EDITOR AND DESIGNER: Sheila Volpe
ASSISTANT ART DIRECTOR: Penny Cobb
PHOTOGRAPHER: Stephen R. Steinhauser
ILLUSTRATOR: Brian Hargreaves
MAPS AND SYMBOLS: Kuo Kang Chen,
Kay Sutaria-Vakil
COPY EDITOR: Jude Ledger
PROOFREADER: Anne Plume
INDEXER: Rosemary Anderson

ART DIRECTOR: Moira Clinch
PUBLISHER: Piers Spence

Manufactured by Universal Graphics,
Singapore
Printed by Midas Printing International
limited, Hong Kong, printed in China

9 8 7 6 5 4 3 2 1

contents

introduction

LEPIDOPTEROLOGY, OR THE study of butterflies, is becoming increasingly popular among the general public, just as many species of these lovely and essentially harmless insects are coming close to extinction. Butterflies are not only fascinating in their own right, but are also extremely accurate indicators of the state of the environment, the degradation of which is a cause for international concern.

There are, of course, some butterflies that will flourish even when their habitat changes, but the majority are set in their evolutionary ways, associated with special habitat requirements, and unable to adapt to radical environmental change.

It is not only exotic and rainforest species that are endangered and little known, but also those butterflies we find on our doorstep that are in need of both research and protection. But the survival of habitats is the key issue for the long-term conservation of all butterflies; maintaining the flora and habitat structure is essential if the battle to protect them is to be won.

Habitat destruction is not the only threat to butterfly populations; certain collectors are also to blame, with highly prized specimens changing hands for substantial sums. It is therefore not enough for species to be recognized as endangered, but action should be taken. Therefore the classification system used here is based on the Convention of International Trade in Endangered Species of Wild Fauna, CITES, which regulates trade in endangered species.

This book is a concise guide to around 700 butterflies ranging from rare to the most common—a small fraction of the 18,000-26,000 butterfly species known. Balanced coverage pays due attention to the five butterfly families, covering their basic features, size, habitat, and status. All six of the internationally recognized faunal regions are covered, embracing the importance of both rare and exotic species and those that we take for granted as an integral part of life around us.

The five butterfly families are arranged according to the most up-to-date taxonomic conventions. We begin with the most colorful, the *Papilionidae*—the swallowtails, then come the *Pieridae*—the whites and sulfurs, followed by the *Nymphalidae*—the brush-footed butterflies, then the *Lycaenidae*— a family made up of coppers, hairstreaks, and blues, and finally the *Riodinidae*—metalmarks.

PAPILIONIDAE: Swallowtails are found worldwide. They are generally large and colorful, and usually characterized by their tails, although in many species, including festoons and apollos, the tails are absent.

PIERIDAE: Whites and sulfurs tend to be easy to identify by their bright colors. Whites are a large and widespread group, while the sulfurs include many of the clouded yellows.

NYMPHALIDAE: Brush-footed butterflies include groups such as the browns, milkweeds, snouts, and aristocrats, all of which have previously been classified as separate families. They are combined here due to their over-riding characteristic: four functional legs.

LYCAENIDAE: This family contains three main groups of butterflies, the hairstreaks, coppers, and blues, each identified by various external characteristics. Butterflies belonging to this family are fairly small.

RIODINIDAE: The common name of this family, metalmarks, refers to the bright metallic spots on the wings of many of its members. Some sources consider this family to be a subfamily of Lycaenidae.

FAMILIES

In the field butterflies can be readily sorted into the five major families. Inspection of their wing shape, venation, and number of legs gives some vital clues. Throughout this book each individual entry has a stylized shape symbol to denote family membership as follows:

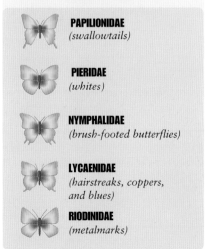

PAPILIONIDAE
(swallowtails)

PIERIDAE
(whites)

NYMPHALIDAE
(brush-footed butterflies)

LYCAENIDAE
(hairstreaks, coppers, and blues)

RIODINIDAE
(metalmarks)

HOW TO USE THIS BOOK

The main part of this book is a directory covering over 700 of the most common species of butterfly found worldwide. Each species is clearly defined by family and genus, and illustrated with a detailed color photograph. Accompanying each entry are symbols indicating size, zone of origin, and conservation status. These symbols are explained here, and there is a handy pull-out key on page 255 for quick reference. In addition each species has a text entry giving information on habitat, ecology, and other characteristics.

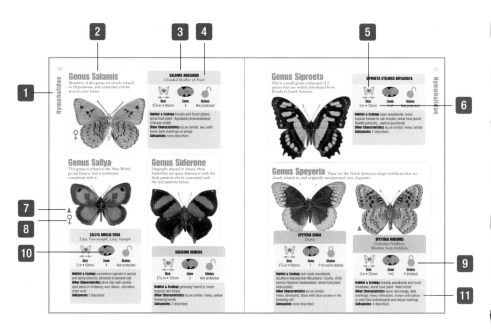

1 AND 5 FAMILY INDICATOR

Each of the five butterfly families is included. The heading at the top of every page indicates to which of the five families the specimens belong.

In addition, each entry is accompanied by a stylized shape symbol showing family membership (see previous page).

2 INTRODUCTION TO GENERA

Each genus has an introduction that outlines some of the typical characteristics of the species within the genus.

Species within the same genus are unified by a matching colored panel behind the butterfly's name—showing, at a glance, which belong to the same genus.

3 SCIENTIFIC NAME

Scientific Latin names are always in a state of flux; the most up-to-date classification is used here. The genus name is followed by the species name and, where appropriate, the subspecies. A butterfly's scientific name reflects something about its morphology, structure, behavior, coloration, and even food plants.

4 COMMON NAME

Where a common name (or in some cases names) is known, it is given underneath the scientific name. Common names can be a source of confusion because different species can have the same common names.

6 SIZE

The sizes given in this book are for the wingspan. This is the distance from the tips of the outstretched forewings. This is a little larger than twice the forewing length since it includes some of the width of the thorax, which may be ¼ in (6mm) in some species. The figure given is for the maximum size of the female, as the female is usually larger than the male.

Convention dictates that the forewing length is taken as a measure of wingspan, but there are numerous tropical butterflies where the hindwing is much longer than the forewing, such as *Papilio androcles* that come from Southeast Asia, and the various Mistletoe hairstreaks, *Iolaus* species, from tropical Africa.

The butterflies are not shown to scale, as showing them in proportion to each other would mean that the smallest specimens would be so small as to be unidentifiable.

7 VENTRAL OR DORSAL VIEW

Most of the specimens have been photographed to show their dorsal (above) surface. However, in some cases the ventral (below) surface is most colorful and is more useful for identification purposes. In cases where the ventral surface is shown, the triangular symbol appears in the frame with the specimen.

8 MALE OR FEMALE SPECIMEN

Most of the specimens shown are male as they are usually more attractive than the corresponding females. However, in those cases where a female specimen is shown, then a female symbol appears in the frame with the specimen.

9 CONSERVATION STATUS

All butterflies that are protected by individual countries, or by conservation groups, are given a closed padlock symbol. If the species has a CITES (Convention of International Trade in Endangered Species of Wild Fauna and Fauna) classification this is also given, CITES I being the highest level of protection, and CITES III being the lowest. We also indicate where species are rare, endangered, or vulnerable. The open padlock symbol indicating an unprotected species represents the status of the butterfly as it is known at the present time.

10 ZONE OF ORIGIN

For purposes of classification the world has been divided up into six faunal regions, which are followed in this book. The map symbol indicates the butterfly's zone of origin (see page 10).

11 TEXT ENTRY

Each entry in the directory section follows the same format:
Habitat and ecology: describes the environmental location of the species. Where the food plants of individual species are known, these are mentioned in the text entry concerned, and where butterflies use numerous food plants these are referred to by the name of the plant family or families.
Other characteristics: the surface of the butterfly not shown in the photograph is detailed, along with a description of the female (or male, on the occasions where the photograph is of the female).
Subspecies: lists the number of subspecies that have been described.

Zone of Origin

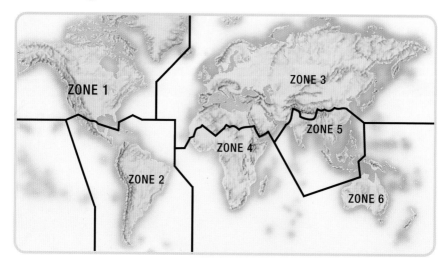

For classification the world has been divided up according to the different types of fauna found in each area.

The key to these regions is as follows:

ZONE 1
Nearctic region

ZONE 2
Neotropical region

ZONE 3
Palearctic region

ZONE 4
African region

ZONE 5
Oriental region

ZONE 6
Australian region

Migration

Technically, migration is the movement from one place to another, and back again. Butterflies never actually do this, instead, if they move at all, it is just one way.

In temperate climates butterflies have to migrate in order to exploit new resources as the seasons progress. In the tropics it is necessary to establish new colonies and territories; if all butterflies stayed where they emerged, mated, and laid eggs, the caterpillars would have too much competition and would very likely starve to death.

There are many stages in between migrant and resident butterflies: there are strong regular migrants, weak irregular or occasional migrants, and species that undergo what is called local movement. "Vagrant" is another term that is used, implying that the butterfly is a migrant but that it turns up only irregularly from a long way away.

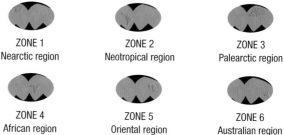

One great migrant, and the major migrant in North America, is the Monarch or Milkweed, *Danaus plexippus*, which makes a return migration to its winter roosts from the Canadian border to the south-west of the United States.

Morphology

The butterfly body is made up of three parts: head, thorax, and abdomen. The head has a pair of antennae or feelers that are usually long and knobbed or clubbed at the ends. The antennae are sensitive to touch and smell, and have a specific number of segments, sometimes of use in identification. There is a pair of compound eyes, on either side of the head. The eyes are bevelled so that a wide angle of vision is possible. The other main feature on the head is the tongue, or proboscis, used for sucking up liquids. Its structure is like two straws fused together and zipped up.

THE THORAX

The thorax is a muscle box with three segments. The three pairs of jointed legs arise from the thorax, one from each of the three segments. In some butterflies the front legs are reduced and non-functional.

THE ABDOMEN

The abdomen contains the bulk of the digestive system, as well as the excretory system. At the tip of the abdomen are the sexual apparatus, called the genitalia, whose internal characteristics can be useful in identifying different species.

THE OUTER BODY

The outer body or integument of the butterfly is covered with small, sensory hairs. Butterflies also have specialized scales on their wings that contain highly volatile insect hormones, called pheromones. These hormones are released into the air during mating, and affect the behavior of the opposite sex.

THE HEAD

The head carries a great deal of sensory apparatus for the butterfly. The largest features are the compound eyes that are made up of thousands of individual eyes, each with a tiny lens and a tiny fraction of view. The bevelled nature of the compound eye means that the butterfly is aware of its immediate environment through a very large angle. The head is covered with minute bristles and hairs that are sensitive to touch, as are the labial palps and the labrum.

This is a typical butterfly in the style of a Swallowtail. Tails when present are on the hindwing. The wings show the veins that support their structure. Three legs are shown, one pair to each segment. The proboscis, or tongue, is normally curled up under the head. Antennae are always clubbed but can be long or short.

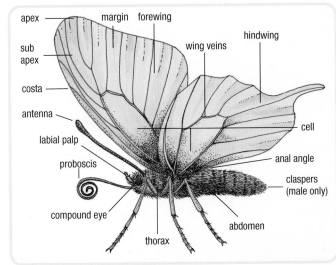

apex — margin forewing

sub apex

hindwing

wing veins

costa

antenna

labial palp

proboscis

compound eye

thorax

cell

anal angle

claspers (male only)

abdomen

CHAPTER 1

Papilionidae

Members of this family are generally large, colorful butterflies with about 500 cosmopolitan species throughout the world. These are the swallowtails characterized by, but not exclusively with, extended tails on the posterior hindwings. There is a wide range of genera in this family including the birdwings or *Ornithoptera* of the southeast Australian region, which are some of the largest butterflies in the world. Other groups include the tailless Parnassians and the festoon butterflies (*Zerynthia*). Some members of the Papilionidae are involved in complicated mimicry complexes with representatives of other butterfly families. These butterflies use a wide variety of larval food plants, and many species are pests on *Citrus* (citrus).

Genus Allancastria

This genus is similar in coloration and pattern to the *Parnassians*. Larval food plants include *Aristolochia* (birthwort).

ALLANCASTRIA CERISY CYPRIA
Eastern Festoon

Size	Zone	Status
2³/₈ in • 60mm	3	Not protected

Habitat & Ecology rocky meadows in montane areas up to 4,000 ft (1,250m)
Other Characteristics BELOW similar;
FEMALE above, more extensive black markings
Subspecies 12 described

Genus Archon

This genus is associated with *Aristolochia* (birthwort) as larvae. Adults faintly resemble the genus *Allancastria* but have distinctive forewing coloration.

ARCHON APOLLINARIS
False Apollo

Size	Zone	Status
2³/₈ in • 60mm	3	Not protected

Habitat & Ecology rocky montane meadows
Other Characteristics BELOW similar;
FEMALE larger with darker markings
Subspecies 2 described

Genus Atrophaneura

Related to *Troides*, many of this genus are models in mimicry complexes. Hindwing tails may be present or absent in males.

ATROPHANEURA HAGENI

Size	Zone	Status
6³/₈ in • 160mm	5	Not protected

Habitat & Ecology montane forests
Other Characteristics BELOW red abdomen;
FEMALE lacks shiny hindwing androconial patch
Subspecies none described

♀

ATROPHANEURA JOPHON
Ceylon Rose

Size	Zone	Status
5^1/$_8$ in • 130mm	5	CITES II

Habitat & Ecology primary rain forest; little known about ecology and life history
Other Characteristics BELOW enlarged reddish marginal spot band; MALE white markings less extensive on both wings
Subspecies none described

ATROPHANEURA NEPTUNUS
Yellow-bodied Club-tail,
Yellow Club-tail

Size	Zone	Status
5^3/$_4$ in • 140mm	5	Not protected

Habitat & Ecology rain forest; low open woodland; larval food plant: *Thottea* (member of birthwort family)
Other Characteristics BELOW reddish white spots along hindwing; FEMALE lacks shiny androconial patch along hindwing inner margin
Subspecies 8 described

ATROPHANEURA SEMPERI SUPERNOTATUS

Size	Zone	Status
5^3/$_4$ in • 140mm	5	Not protected

Habitat & Ecology Indonesian rain forest
Other Characteristics BELOW paler; red postmedian and submarginal spot bands; FEMALE duller; lacks androconial patch
Subspecies 8 described

Genus Baronia

This monotypic genus is endemic to Mexico and possesses primitive features. Its larvae are associated with *Acacia cymbispina* (a member of the legume family*)*.

BARONIA BREVICORNIS

Size	Zone	Status
2³/₈ in • 60mm	2	Rare

Habitat & Ecology moist meadows to semi-desert conditions
Other Characteristics BELOW similar; FEMALE larger; usually more yellow or orange
Subspecies 2 described

Genus Battus

This genus comprises 16 species, which range widely in North and South America. Larvae are associated with the *Aristolochiaceae* (birthwort family).

♀

BATTUS ZETIDES
Zetides Swallowtail

Size	Zone	Status
1³/₄ in • 40mm	2	CITES I

Habitat & Ecology rain forest canopies
Other Characteristics BELOW black hindwing with silver rays; submarginal spots; golden yellow along submargin; MALE similar
Subspecies none described

BATTUS POLYDAMAS LUCIANUS
Polydamas Swallowtail, Gold Rim

Size	Zone	Status
4 in • 100mm	1•2	Not protected

Habitat & Ecology varies from xeric woodland to tropical forest
Other Characteristics BELOW red submarginal spot band outlined in greenish yellow on hindwing; FEMALE lacks androconial fold
Subspecies 14 described

♀

Genus Bhutanitis

The complex wing patterns and coloration make this genus distinctive, however, little is known about the ecology and biology.

BHUTANITIS MANSFIELDI
Mansfield's Three-tailed Swallowtail

Size	Zone	Status
3 1/8 in • 80mm	5	CITES II

Habitat & Ecology montane forests; little known about ecology and life history
Other Characteristics BELOW paler; MALE similar
Subspecies 2 described

Genus Eleppone

This Australian monotypic genus has only recently been described. It is associated with *Citrus* (citrus) as larvae.

Genus Cressida

This monotypic genus is endemic to Australia. The claspers (*valvae*) are absent in the males.

CRESSIDA CRESSIDA

Size	Zone	Status
4 in • 100mm	6	Not protected

Habitat & Ecology coastal forest; woodland up to 600 ft (200m); larval food plant: *Aristolochia* (birthwort)
Other Characteristics BELOW similar;
FEMALE forewings more transparent
Subspecies 5 described

ELEPPONE ANACTUS
Dingy Swallowtail

Size	Zone	Status
2 3/8 in • 60mm	6	Not protected

Habitat & Ecology open woodland and groves; pest of *Citrus* (citrus)
Other Characteristics BELOW similar;
FEMALE similar
Subspecies none described

Genus Eurytides

This genus is commonly called the "kite swallowtails" because of their elongated tails. They are excellent fliers, very common where they occur, and often observed at mud puddles. This genus currently contains nearly 50 species, and the larvae are associated with various members of the *Annonaceae* (custard apple family).

EURYTIDES CELADON
Cuban Kite Swallowtail

Size	Zone	Status
3¹/₈ in • 80mm	1•2	Not protected

Habitat & Ecology open woodland and mesic forest edges; life history unknown
Other Characteristics BELOW pale brown; red suffusion along hindwing; blue spot near anal angle; FEMALE above, markings blue green
Subspecies none described

EURYTIDES BELLEROPHON

Size	Zone	Status
4 in • 100mm	2	Not protected

Habitat & Ecology open woodland; fragmented forest along streams; montane cloud forest
Other Characteristics BELOW similar; FEMALE similar
Subspecies none described

♀

EURYTIDES IPHITAS
Yellow Kite

Size	Zone	Status
3⁷/₈ in • 95mm	2	Vulnerable

Habitat & Ecology canyons; coastal mountains
Other Characteristics BELOW darker hindwing veins; black transverse median band; silvery blue marginal band; MALE similar
Subspecies none described

EURYTIDES L. LYSITHOUS
Harris' Mimic Swallowtail

Size	**Zone**	**Status**
3$\frac{1}{8}$ in • 80mm	2	Vulnerable

Habitat & Ecology Atlantic rain forest; larval food plant: *Annona acutiflora* (custard apple family)
Other Characteristics BELOW lacks red submarginal spot band; FEMALE similar
Subspecies 2 described

EURYTIDES MARCELLUS
Zebra swallowtail

Size	**Zone**	**Status**
3$\frac{5}{8}$ in • 90mm	1	Vulnerable

Habitat & Ecology moist low woodlands near swamps and rivers; adults frequent nectar plants
Other Characteristics BELOW angular red postmedian band on hindwing; FEMALE similar
Subspecies none described

EURYTIDES PAUSANIAS

Size	**Zone**	**Status**
3$\frac{5}{8}$ in • 90mm	2	Protected

Habitat & Ecology open areas; rain forest edges; little known about life history
Other Characteristics BELOW brown black ground color; red basal rays; red submarginal chevrons on hindwing; FEMALE similar
Subspecies 4 described

Genus Graphium

This genus comprises more than 100 species from Africa, Europe, and Asia, and the long tails may be either present or absent. These butterflies occur in a wide variety of habitats and can often be observed at mud puddles.

GRAPHIUM ANDROCLES

Size	Zone	Status
3⅝ in • 90mm	5	Not protected

Habitat & Ecology dense forest; mud puddles
Other Characteristics BELOW paler; golden suffusion at base of wings, along submargin and near anal angle; FEMALE similar
Subspecies 3 described

GRAPHIUM AGAMEMNON
Tailed Jay,
Green-spotted Triangle Butterfly

Size	Zone	Status
4 in • 100mm	5•6	Not protected

Habitat & Ecology open woodland; forest edges
Other Characteristics BELOW paler black; blue eyespot near anterior hindwing margin; FEMALE broader wings
Subspecies more than 20 described

GRAPHIUM ANGOLANUS
Angola White, Lady Swallowtail,
White Lady Swallowtail

Size	Zone	Status
3⅛ in • 80mm	4	Not protected

Habitat & Ecology warm savannahs; avid flower visitors; often seen at mud puddles
Other Characteristics BELOW ochreous ground color on forewing apex and hindwing margin; FEMALE forewing margin less concave
Subspecies 3 described

GRAPHIUM CODRUS

Size	Zone	Status
$3^7/_8$ in • 95mm	5•6	Not protected

Habitat & Ecology rain forest; secondary forests; along coastal beach strands
Other Characteristics BELOW reduced white patch on hindwing; FEMALE above, lacks grayish white androconial patches of hindwing
Subspecies 16 described

GRAPHIUM COLONNA
Mamba, Black Swordtail

Size	Zone	Status
$3^1/_8$ in • 80mm	4	Not protected

Habitat & Ecology wooded areas; larval food plant: *Annonaceae* (custard apple family)
Other Characteristics BELOW paler with reddish postmedian spot band; FEMALE similar
Subspecies none described

GRAPHIUM DELESSERTI
Zebra, Malayan Zebra

Size	Zone	Status
$3^5/_8$ in • 90mm	5	Not protected

Habitat & Ecology open forests and woods; males are attracted to mud puddles
Other Characteristics BELOW larger golden orange patch on hindwing; FEMALE similar
Subspecies 3 described

GRAPHIUM EPAMINONDAS

Size	Zone	Status
3³/₈ in • 85mm	5	Rare

Habitat & Ecology lowland rain forest; restricted and protected in the Andaman Islands
Other Characteristics ABOVE lacks yellow suffusion; FEMALE above, black and white
Subspecies 2 described

GRAPHIUM GELON

Size	Zone	Status
2³/₈ in • 60mm	6	Not protected

Habitat & Ecology forests, restricted to the Loyalty Islands and New Caledonia
Other Characteristics BELOW warm brown ground color; paler markings; FEMALE above, brown, with prominent greenish bands
Subspecies none described

GRAPHIUM IDAEOIDES

Size	Zone	Status
5¹/₈ in • 130mm	5	Protected

Habitat & Ecology rain forest clearings; along stream edges; perfect mimic of *Idea leucone*
Other Characteristics ABOVE similar; FEMALE similar
Subspecies none described

GRAPHIUM MENDANA

Size
4⅛ in • 105mm

Zone
6

Status
Rare

Habitat & Ecology lowland rain forest
Other Characteristics BELOW brown ground color; yellow submarginal spot band on forewing; red spots along costa, and inner margin of hindwing; FEMALE similar
Subspecies 4 described

GRAPHIUM MORANIA
White Lady Swallowtail

Size
2⅞ in • 70mm

Zone
4

Status
Not protected

Habitat & Ecology savannah; forest edges
Other Characteristics BELOW pale brownish black suffused with red at forewing base; overscaled with rust at forewing apex and along lateral hindwing margin; FEMALE similar
Subspecies none described

GRAPHIUM POLICENES
Common Swordtail,
Small Striped Swordtail

Size
3⅛ in • 80mm

Zone
4

Status
Not protected

Habitat & Ecology mesic or xeric woods
Other Characteristics BELOW paler; expanded markings; basal hindwing band suffused with red; two brown spots on hindwing; FEMALE similar
Subspecies 3 described

GRAPHIUM SANDAWANUM

Size | **Zone** | **Status**
2⁷/₈ in • 70mm | 5 | Vulnerable

Habitat & Ecology humid Philippine rain forest
Other Characteristics ABOVE black, with more vivid blue green areas; FEMALE similar
Subspecies none described

GRAPHIUM THULE

Size | **Zone** | **Status**
2⁷/₈ in • 70mm | 6 | Not protected

Habitat & Ecology localized species in rain forest and marginal secondary forests of New Guinea and Irian Jaya
Other Characteristics BELOW paler; enlarged submarginal spot band; FEMALE similar
Subspecies 2 described

GRAPHIUM STRESEMANNI

Size | **Zone** | **Status**
3¹/₈ in • 80mm | 6 | Rare

Habitat & Ecology rain forest
Other Characteristics BELOW paler brown ground color; emerald green markings; FEMALE duller
Subspecies none described

GRAPHIUM WEISKEI
Purple Spotted Swallowtail

Size	Zone	Status
3¹/₈ in • 80mm	6	Rare

Habitat & Ecology canopy of highland rain forest
Other Characteristics BELOW overscaled with white at forewing apex; green markings instead of lavender; emerald green at wing bases and on hindwing; FEMALE similar
Subspecies 2 described

Genus Luehdorfia

This is a small genus comprising 4 species, 3 of which are listed in the Red Data book on Swallowtails. There is little further information available on this group.

Genus Iphiclides

This genus comprises 2 species, and is restricted to the temperate zone of Eurasia.

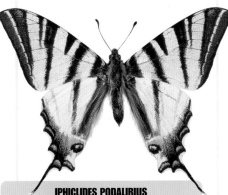

IPHICLIDES PODALIRIUS
Scarce Swallowtail

Size	Zone	Status
3⁵/₈ in • 90mm	3	Not protected

Habitat & Ecology paths; hedges; waysides; larval food plant: *Prunus* (plum family)
Other Characteristics BELOW paler; FEMALE lacks gray androconial patches
Subspecies 8 described

LUEHDORFIA CHINENSIS

Size	Zone	Status
2³/₈ in • 60mm	3	Rare

Habitat & Ecology flies in open woodlands
Other Characteristics ABOVE paler; FEMALE similar
Subspecies 2 described

Genus Meandrusa

In the males of this Asian genus comprising 2 species, androconia are absent and the forewing apices are particularly recurved.

MEANDRUSA PAYENI EVAN
Yellow Gorgon, Outlet Sword, Sickle

Size	Zone	Status
4⁷/₈ in • 120mm	5	Vulnerable

Habitat & Ecology rain forest
Other Characteristics BELOW dark rust-colored markings; silvery spots near inner hindwing margin; FEMALE less acute forewing apex; lacks dark brown markings
Subspecies 7 described

Genus Ornithoptera

All 13 species are extremely large, and noted for their pronounced sexual dimorphism. The male wings include an array of bright iridescent green, orange, gold, or blue, which contrast with the dark brown or black ground color. The females are markedly sexually dimorphic, usually a duller brownish black to black with white to cream spots, and/or whitish yellow patches on the wings.

ORNITHOPTERA ALEXANDRAE
Queen Alexandra's Birdwing

Size	Zone	Status
11¹/₈ in • 280mm	6	CITES I

Habitat & Ecology rain forest canopy; secondary forest; larval food plant: *Aristolochiaceae* (birthwort family)
Other Characteristics BELOW black overscaled with greenish gold along anterior of wings: FEMALE black; white spots and grayish white patches on hindwing
Subspecies none described

ORNITHOPTERA CHIMAERA
Chimaera Birdwing

Size	Zone	Status
7⁵/₈ in • 190mm	6	CITES II

Habitat & Ecology montane rain forest
Other Characteristics BELOW forewing overscaled with chartreuse; black postmedian band; FEMALE brownish black ground color; white spot band on forewing; below, brighter
Subspecies 2 described

ORNITHOPTERA CROESUS LYDIUS

| **Size** | **Zone** | **Status** |
| 7⁵/₈ in • 190mm | 6 | CITES II |

Habitat & Ecology rain forest
Other Characteristics BELOW similar;
FEMALE blackish brown ground color; cream spots at apex; dull yellow postmedian submarginal band
Subspecies 4 described

ORNITHOPTERA GOLIATH SAMSON
Goliath Birdwing

| **Size** | **Zone** | **Status** |
| 8³/₈ in • 210mm | 6 | CITES II |

Habitat & Ecology primary and secondary rain forest; takes nectar at flowering trees
Other Characteristics BELOW golden green overscaling on forewing cell and lower part of wing; FEMALE blackish brown with pale yellow markings on forewing; hindwing patch golden
Subspecies 5 described

ORNITHOPTERA MERIDIONALIS

| **Size** | **Zone** | **Status** |
| 6 in • 150mm | 6 | CITES II |

Habitat & Ecology low and montane rain forest
Other Characteristics BELOW darkened veins; iridescent green on basal forewing cell; FEMALE blackish brown ground color; cream patches in forewing cell, postmedian area, apex, and along submargin; cream hindwing margin shades to dull gold; prominent black postmedian spot band
Subspecies 2 described

ORNITHOPTERA PARADISEA ARFAKENSIS
Paradise Birdwing, Tailed Birdwing

Size	**Zone**	**Status**
$6^3/8$ in • 160mm	6	CITES II

Habitat & Ecology rain forest; populations local
Other Characteristics BELOW forewing overscaled with green, shading to gold; FEMALE blackish brown; cream spots in cell, apex, and submargin; black postmedian spot band on hindwing
Subspecies 5 described

ORNITHOPTERA PRIAMUS
Priam's Birdwing

Size	**Zone**	**Status**
$6^5/8$ in • 130mm	6	CITES II

Habitat & Ecology marginal secondary forest; larval food plant: *A. tagala* (Indian birthwort)
Other Characteristics BELOW forewing overscaled with green; FEMALE black brown ground color; cream patches; black postmedian spot band
Subspecies 14 described

ORNITHOPTERA ROTHSCHILDI
Rothschild's Birdwing

Size	**Zone**	**Status**
6 in • 150mm	6	CITES II

Habitat & Ecology highland rain forest
Other Characteristics BELOW overscaled with green gold to margin; FEMALE blackish brown; duller; cream spots on hindwing
Subspecies none described

ORNITHOPTERA TITHONUS

Size	Zone	Status
6 in • 150mm	6	CITES II

Habitat & Ecology highland rain forest
Other Characteristics BELOW gold in cell; darker veins; black submarginal spot band; FEMALE black ground color; cream spots on forewing; cream/gold patch on hindwing
Subspecies 4 described

ORNITHOPTERA VICTORIAE
Queen Victoria's Birdwing

Size	Zone	Status
6⁷/₈ in • 170mm	6	CITES II

Habitat & Ecology low secondary forest; gardens
Other Characteristics BELOW lacks black on inner margin; darker cell veins; black spots on lateral margin; FEMALE black; pale cream patches in cell, subapex, at apex; below, yellow spots at base of wings; submarginal spot band
Subspecies 7 described

Genus Pachliopta

With more than a dozen species from the Indian and Australian regions, this genus is sometimes merged with the closely related *Atrophaneura*. Most species are closely associated with the more toxic members of the *Aristolochiaceae* (birthwort family) as larval food plants, and the larvae concentrate these poisons for protection from predators. As adults, they advertise their distastefulness by the prominent red patches on the wings and the body.

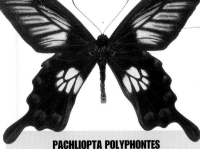

PACHLIOPTA POLYPHONTES

Size	Zone	Status
3¹/₈ in • 80mm	6	Not protected

Habitat & Ecology forest
Other Characteristics BELOW similar; FEMALE duller
Subspecies 5 described

Genus Papilio

With more than 200 species described, this group is one of the most widespread genera, and a particular favorite of collectors. Most of these butterflies are associated with the tropical climes, and the majority have long tails. The caterpillars exploit members of the *Rutaceae* (rue family), *Lauraceae* (laurel family), and *Umbelliferae* (apiaceae family) for food, and *Citrus* (citrus) is often used as a host plant for many species.

PAPILIO AMYNTHOR

	Size	Zone	Status
	4¹/₂ in • 115mm	6	Not protected

Habitat & Ecology orchards; citrus groves; commonly found
Other Characteristics BELOW similar; FEMALE similar
Subspecies 2 described

PAPILIO ALEXANOR
Southern Swallowtail

	Size	Zone	Status
	2⁵/₈ in • 65mm	3	Protected

Habitat & Ecology alpine meadows; thistle visitor; larval food plant: *Umbelliferae* (apiaceae family)
Other Characteristics BELOW similar; FEMALE wing color and pattern are reminiscent of *P. machaon* and *P. hospiton*
Subspecies 6 described

PAPILIO ANCHISIADES
Red-spotted Swallowtail

	Size	Zone	Status
	3⁷/₈ in • 95mm	2	Not protected

Habitat & Ecology open woodland; forest; disturbed areas
Other Characteristics BELOW similar, but less pink; FEMALE lighter forewings; above and below, 2 reddish pink spot bands
Subspecies 4 described

PAPILIO ANDROGEUS
Queen Page

Size	**Zone**	**Status**
4^1/$_8$ in • 105mm	1•2	Not protected

Habitat & Ecology rain forest; tropical dry forest
Other Characteristics sexes markedly dimorphic; BELOW similar; FEMALE black with a subtle blue fusion on the wings; lacks tail
Subspecies 5 described

PAPILIO ANTIMACHUS
African Giant Swallowtail

Size	**Zone**	**Status**
9^1/$_8$ in • 230mm	4	Not protected

Habitat & Ecology tropical forest; males patrol along the forest edges or aggregate along mud puddles or near streams; females are more reclusive and associated with forest canopy
Other Characteristics ABOVE dull orange with black markings; FEMALE smaller than male by up to 4 in (100mm); forewing apex is more rounded
Subspecies 3 described

PAPILIO ARISTODEMUS PONCEANUS
Schaus' Swallowtail

Size	**Zone**	**Status**
3^7/$_8$ in • 95mm	1•2	Protected

Habitat & Ecology tropical hammocks, often seen flying along paths through these areas; larval food plant: *Amyris* (torchwood), occasionally associated with *Xanthoxylum* (evodia)
Other Characteristics ABOVE ground color blackish brown; yellow postmedian/median band on both wings; submarginal spot bands; FEMALE similar
Subspecies 5-6 described

PAPILIO ARISTOR
Scarce Haitian Swallowtail

Size
4 in • 100mm

Zone
2

Status
Rare

Habitat & Ecology dry scrub or mesic woodlands (Hispaniola); little known about life history or ecology
Other Characteristics BELOW similar; FEMALE similar
Subspecies none described

PAPILIO ASCOLIUS

Size
4³/₈ in • 110mm

Zone
2

Status
Not protected

Habitat & Ecology humid tropical rain forest; males in canopy; uncommon
Other Characteristics BELOW has white postmedian spot band; FEMALE orange on base of both wings
Subspecies 5 described

PAPILIO BENQUETANA

Size
4 in • 100mm

Zone
5

Status
Rare

Habitat & Ecology restricted to a few forests in the Philippines
Other Characteristics BELOW black submarginal spot band overlaid with yellow; FEMALE similar
Subspecies none described

PAPILIO BLUMEI

Size	Zone	Status
4⁷/₈ in • 120mm	6	Not protected

Habitat & Ecology common in rain forest
Other Characteristics BELOW veins darkened on forewing; orange submarginal spot band edged in blue on hindwing; FEMALE blue green bands
Subspecies 4 described

PAPILIO BREVICAUDA BRETONENSIS
Cape Breton Swallowtail,
Maritime Swallowtail,
Short-tailed Swallowtail

Size	Zone	Status
3¹/₈ in • 80mm	1	Not protected

Habitat & Ecology open woodland; paths; scrub areas; common
Other Characteristics ABOVE yellow spot bands on both wings; MALE very similar
Subspecies 3 described

PAPILIO CAIGUANABUS
Poey's Black Swallowtail

Size	Zone	Status
4 in • 100mm	2	Rare

Habitat & Ecology localized; endemic to the mesic tropical forests of Cuba
Other Characteristics ABOVE white hindwing spot band; MALE above, blackish brown with golden yellow forewing submarginal spot band on both wings
Subspecies none described

PAPILIO CANOPUS
Canopus Butterfly

Size	Zone	Status
3⁵/₈ in • 90mm	6	Not protected

Habitat & Ecology very local; endemic to rain forest of Indonesia and northern Australia
Other Characteristics BELOW orange on tornus; FEMALE similar
Subspecies more than 12 described

PAPILIO CLYTIA
Common Mime

Size	Zone	Status
4⁷/₈ in • 120mm	5	Protected

Habitat & Ecology woodland
Other Characteristics both sexes have several color forms, ranging from yellowish orange to yellow green, to a dark smoky brown; BELOW has darker veins; orange on hindwing submargin; FEMALE similar
Subspecies 9 described

PAPILIO CHIKAE
Luzon Peacock Swallowtail

Size	Zone	Status
5¹/₈ in • 130mm	5	CITES I

Habitat & Ecology rain forest of the Philippines
Other Characteristics BELOW fuchsia spot band; FEMALE similar
Subspecies none described

PAPILIO CONSTANTINUS
Constantine's Swallowtail

Size	Zone	Status
4³/₈ in • 110mm	4	Not protected

Habitat & Ecology mesic scrubby woodlands; forest edges
Other Characteristics BELOW has numerous yellow rayed markings; FEMALE slightly larger; rounder forewings
Subspecies 3 described

PAPILIO CRESPHONTES
Giant Swallowtail, Orange Dog

Family & Size	Zone	Status
5³/₄ in • 140mm	1•2	Not protected

Habitat & Ecology old woodland and forest; larval food plant: *Citrus* (citrus)
Other Characteristics BELOW ground color yellow; has blue azure postmedian spot band on hindwing; FEMALE similar
Subspecies none described

PAPILIO CRINO
Common Banded Peacock

Size	Zone	Status
4 in • 100mm	5	Not protected

Habitat & Ecology dry woodland; secondary forest; gardens; avid flower visitor
Other Characteristics BELOW grayish white forewing postmedian band; FEMALE larger
Subspecies none described

PAPILIO DARDANUS
Mocker Swallowtail

Size
4³/₈ in • 110mm

Zone
4

Status
Not protected

Habitat & Ecology common and widespread in woods and citrus groves
Other Characteristics BELOW similar;
FEMALE highly variable, can be combinations of black, white, orange, or yellow
Subspecies 12 described

PAPILIO DEMODOCUS
Citrus Butterfly, African Lime Butterfly,
Christmas Butterfly, Orange Dog

Size
4¹/₂ in • 115mm

Zone
4

Status
Not protected

Habitat & Ecology open woodland; forests; citrus groves; can be considered a pest
Other Characteristics BELOW similar;
FEMALE similar
Subspecies none described

PAPILIO ESPERANZA

Size
4³/₈ in • 110mm

Zone
2

Status
Rare

Habitat & Ecology fragmented forest
Other Characteristics BELOW black postmedian spot band on hindwing; iridescent bluish scales;
FEMALE similar
Subspecies none described

PAPILIO EUCHENOR

Size	Zone	Status
2⁷/₈ in • 70mm	6	Not protected

Habitat & Ecology secondary forest edges
Other Characteristics ABOVE yellow rays in forewing cell; blue eyespot along hindwing costa; FEMALE larger
Subspecies 13 described

PAPILIO FUSCUS
Yellow Helen, Black and White Helen,
Banded Helen

Size	Zone	Status
5³/₄ in • 140mm	5•6	Not protected

Habitat & Ecology secondary forest; gardens; very abundant; larval food plant: *Rutaceae* (rue family)
Other Characteristics ABOVE pale yellow on anterior hindwing; FEMALE blackish brown ground color
Subspecies more than 20 described

PAPILIO GAMBRISIUS BURUANUS

Size	Zone	Status
6³/₈ in • 160mm	5	Not protected

Habitat & Ecology rain forest
Other Characteristics BELOW blue postmedian spot band on hindwing; FEMALE similar
Subspecies 2 described

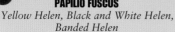

PAPILIO GARLEPPI

Size	Zone	Status
4 in • 100mm	2	Rare

Habitat & Ecology rain forest; along forest edges; little known about ecology and life history
Other Characteristics BELOW similar; FEMALE dimorphic; black with blue and red submarginal markings on hindwing
Subspecies 3 described

PAPILIO GLAUCUS
Tiger Swallowtail

Size	Zone	Status
4¹/₈ in • 105mm	1	Not protected

Habitat & Ecology open woodland; gardens, forest edges; uncommon in Washington State (end of range), where it is of special concern
Other Characteristics BELOW similar; FEMALE more blue on hindwing margin, with black morph; below, blackish brown; prominent tiger stripes
Subspecies 4 of questionable validity described

PAPILIO HECTORIDES

Size	Zone	Status
5¹/₈ in • 130mm	2	Not protected

Habitat & Ecology open areas in dense woods; along forest edges; mimics *Parides* models
Other Characteristics BELOW orange postmedian band on hindwing; FEMALE dimorphic; black with similar diagonal white band; white transverse band on the forewing; more red spots on hindwing disk
Subspecies 3 described

PAPILIO HOMERUS
Homerus Swallowtail

Size	Zone	Status
6 in • 150mm	2	CITES

Habitat & Ecology fragmented tropical forest
Other Characteristics ABOVE blue postmedian spot band and yellow orange submarginal bars on hindwing; FEMALE enlarged blue spots on postmedian band
Subspecies none described

PAPILIO HORNIMANI
Horniman's Swallowtail

Size	Zone	Status
4 in • 100mm	4	Not protected

Habitat & Ecology montane forests
Other Characteristics BELOW brown with yellow submarginal spots; FEMALE above, blue more subdued; hindwing marginal spots reduced
Subspecies 1 described

PAPILIO HOSPITON
Corsican Swallowtail

Size	Zone	Status
3 in • 75mm	3	CITES I

Habitat & Ecology open montane woodland of Corsica and Sardinia; larval food plant: *Ferula communis* (giant fennel)
Other Characteristics BELOW forewing yellow at base, at apex, and along submargin; yellow orange chevrons on hindwing; FEMALE similar
Subspecies none described

PAPILIO JACKSONI
Jackson's Swallowtail

Size	Zone	Status
3⅝ in • 90mm	4	Not protected

Habitat & Ecology low- to medium-elevation forest
Other Characteristics BELOW rust-colored patterns; forewing overscaled with white at apex; hindwing submarginal band reduced; FEMALE additional white spots at subapex and in forewing cell; yellow patch in cell of hindwing
Subspecies 6 described

PAPILIO JORDANI

Size	Zone	Status
6⅞ in • 170mm	6	Rare

Habitat & Ecology forests of Sulawesi; primary forests near streams; female mimics *Idea blanchardi*; larval food plant: *Rutaceae* (rue family)
Other Characteristics BELOW similar; FEMALE dimorphic; resembles *Idea*
Subspecies none described

PAPILIO LAGLAIZEI

Size	Zone	Status
4 in • 100mm	5•6	Not protected

Habitat & Ecology lowland and montane forest; marginal secondary forest to 4,500 ft (1,500m)
Other Characteristics ABOVE black ground color; diffuse bluish green subapical bar on forewing; median band from costa to tornus to hindwing; FEMALE above, broader and brighter bluish green bands; below, brighter
Subspecies 4 described

PAPILIO LORMIERI
Western Emperor Swallowtail,
Emperor Swallowtail

Size	Zone	Status
5^1/$_8$ in • 130mm	4	Not protected

Habitat & Ecology rain forest; larval food plant:
Citrus (citrus)
Other Characteristics BELOW paler ground color;
yellow markings enlarged; FEMALE similar
Subspecies 3 described

PAPILIO LORQUINIANUS

Size	Zone	Status
4^7/$_8$ in • 120mm	6	Not protected

Habitat & Ecology forests of Indonesia
Other Characteristics BELOW black postmedian
band edged in dull orange on hindwing; outer
margin with iridescent blue; FEMALE similar
Subspecies 5 described

PAPILIO MACHAON
Artemisia Swallowtail,
Old World Swallowtail

Size	Zone	Status
3 in • 75mm	1•3	Not protected

Habitat & Ecology old forest; woodland;
meadows; often in the Arctic
Other Characteristics BELOW overscaled with
yellow on forewing base, hindwing inner margin,
and along submarginal bands of both wings;
FEMALE similar
Subspecies more than 40 described

PAPILIO MARAHO
Broad-tailed Swallowtail

Size	Zone	Status
4⁷/₈ in • 120mm	5	Vulnerable

Habitat & Ecology protected in montane rain forest of Taiwan; larval food plant: *Sassafras randaiensis* (member of laurel family)
Other Characteristics BELOW similar;
FEMALE similar
Subspecies none described

PAPILIO MONTROUZIERI

Size	Zone	Status
3 in • 75mm	6	Not protected

Habitat & Ecology forest and forest edges
Other Characteristics BELOW forewing white near end cell; lighter brown submarginal spot band with lighter dull rust along outer margin of hindwing; FEMALE blue patches duller
Subspecies none described

PAPILIO NEUMOEGENI

Size	Zone	Status
3¹/₈ in • 80mm	5	Vulnerable

Habitat & Ecology rain forest canopies
Other Characteristics BELOW a few iridescent gold scales on basal half of both wings;
FEMALE above, lacks black androconial patch on forewing
Subspecies none described

PAPILIO NIREUS
Green-banded Swallowtail,
Narrow-blue-banded Swallowtail

Size	Zone	Status
4 in • 100mm	4	Not protected

Habitat & Ecology woodland; forests; gardens; groves with *Citrus* (citrus)
Other Characteristics BELOW brown ground color; overscaled with rust on forewing apex, along margins, and hindwing; yellow submarginal spot band on hindwing; FEMALE similar
Subspecies 4 described

PAPILIO OENOMAUS

Size	Zone	Status
6 in • 150mm	5	Not protected

Habitat & Ecology rain forest
Other Characteristics BELOW dark red basal spots on both wings; hindwing submarginal spot band capped with dull pinkish red; FEMALE median band outlined in dull red on forewing
Subspecies 2 described

PAPILIO PALINURUS
Banded Peacock, Burmese Peacock,
Moss Peacock

Size	Zone	Status
3³/₈ in • 85mm	5	Vulnerable

Habitat & Ecology rain forest
Other Characteristics BELOW paler; white apical spot on hindwing; cream band on forewing; FEMALE enlarged orange spots
Subspecies 5 described

PAPILIO PELODURUS
Eastern Black and Yellow Swallowtail

Size	**Zone**	**Status**
4⁷/₈ in • 120mm	4	Not protected

Habitat & Ecology montane forest
Other Characteristics BELOW pale golden brown; lighter submarginal spots on hindwing; FEMALE duller
Subspecies 2 described

PAPILIO PHORBANTA EPIPHORBAS
Papillon La Pature

Size	**Zone**	**Status**
4³/₈ in • 110mm	4	Vulnerable

Habitat & Ecology restricted to Réunion; larval food plant: *Citrus* (citrus)
Other Characteristics BELOW yellow spots on hindwing; FEMALE dimorphic; above, cream markings
Subspecies none described

PAPILIO PILUMNUS
Three-tailed Swallowtail

Size	**Zone**	**Status**
4¹/₈ in • 105mm	1•2	Not protected

Habitat & Ecology woodland scrub; mesic forest; larval food plant: *Litsea glaucescens* (member of the laurel family)
Other Characteristics BELOW paler; larger faint blue submarginal band on hindwing; FEMALE similar
Subspecies none described

PAPILIO POLYMNESTOR
Blue Mormon

Size	**Zone**	**Status**
5¹/₈ in • 130mm	5	Not protected

Habitat & Ecology deep forests; woods; gardens to take nectar
Other Characteristics BELOW dull gray; basal red markings; FEMALE below, tan patch on hindwing
Subspecies 3 described

PAPILIO REX
Regal Swallowtail, King Papilio

Size	**Zone**	**Status**
5¹/₈ in • 130mm	4	Not protected

Habitat & Ecology rain forest; mimics danaids
Other Characteristics BELOW yellow markings expanded; pale rust horizontal markings under forewing cell; FEMALE more rounded wings
Subspecies 8 described

PAPILIO SCHMELTZI

Size	**Zone**	**Status**
4³/₈ in • 110mm	6	Not protected

Habitat & Ecology open clearings; forest edges; open country; endemic to Fiji
Other Characteristics BELOW iridescent blue; postdiscal spot band on hindwing;
FEMALE similar
Subspecies none described

PAPILIO TOBOROI

Size	Zone	Status
4 in • 100mm	6	Threatened

Habitat & Ecology rain forest; marginal secondary forest of the Solomon Islands (including Bougainville)
Other Characteristics ABOVE dull iridescent median blue band extends toward tornus on forewing; hindwing with dull diffuse iridescent blue at base; MALE similar
Subspecies 2 described

PAPILIO VICTORINUS

Size	Zone	Status
4³/₈ in • 110mm	1•2	Not protected

Habitat & Ecology rain forest; mesic forest; woods
Other Characteristics BELOW red postmedian and submarginal spot bands on hindwing; FEMALE reddish markings on the hindwing tornus
Subspecies 3 described

PAPILIO ULYSSES
Ulysses Butterfly, Blue Mountain Swallowtail, Mountain Blue, Blue Emperor

Size	Zone	Status
5³/₄ in • 140mm	6	Protected

Habitat & Ecology rain forest clearings and paths
Other Characteristics BELOW covered with white from apex to tornus; hindwing sprinkled with white; prominent black eyespot at costa and on tornus; FEMALE similar
Subspecies 18 described

PAPILIO WEYMERI

Size	Zone	Status
6 in • 150mm	6	Not protected

Habitat & Ecology rain forest and woods on two of the Admiralty Islands; infrequently seen; larval food plant: *Micromelum minutum* (member of the rue family)
Other Characteristics BELOW lacks dark greenish sheen; postmedian spot band edged in iridescent blue on hindwing; orange spot above tornus; FEMALE paler; duller
Subspecies none described

PAPILIO XANTHOPLEURA

Size	Zone	Status
5 in • 125mm	2	Rare

Habitat & Ecology Amazonian rain forest basin; little known about life history
Other Characteristics BELOW iridescent greenish blue at end cell of forewing band extending to tornus and along submargin; FEMALE dimorphic
Subspecies none described

PAPILIO XUTHUS
Chinese Yellow Swallowtail

Size	Zone	Status
4³/₈ in • 110mm	3•5	Not protected

Habitat & Ecology open woodland; gardens; larval food plant: *Poncris* (member of the citrus family)
Other Characteristics cream, white, and yellow forms in both sexes; BELOW paler; 2 black postdiscal spots on hindwing; FEMALE similar
Subspecies 6 described

Genus Parides

The genus of exclusively neotropical species is characterized by the scalloped hindwing margins without elongated tails, and by the white, woolly androconia contained in a fold on the hindwing inner margin. The larval food plants are members of the *Aristolochiaceae* (birthwort family).

PARIDES GUNDLACHIANUS
Gundlach's Swallowtail

Size	Zone	Status
3^1/$_8$ in • 80mm	2	Not protected

Habitat & Ecology mesic, montane, and tropical forests in coastal calcite canyons
Other Characteristics ABOVE enlarged blue patch; expanded red submarginal spot band on hindwing; FEMALE more extensive red markings; above, lacks white cell on forewing
Subspecies 2 described

PARIDES EURIMEDES MYLOTES
Cattle Heart

Size	Zone	Status
2 in • 50mm	2	Not protected

Habitat & Ecology tropical rain, or mesic forests; often seen at mud puddles
Other Characteristics BELOW paler; lacks forewing patch; red hindwing band reduced; FEMALE white patches on both wings
Subspecies 6 described

PARIDES HAHNELI
Hahneli Amazon Swallowtail

Size	Zone	Status
4 in • 100mm	2	Threatened

Habitat & Ecology Amazonian rain forest, especially along the old sand strips or beaches
Other Characteristics BELOW similar; FEMALE larger
Subspecies none described

PARIDES OROPHOBUS
Ascanius Swallowtail,
Fluminense Swallowtail

Size	**Zone**	**Status**
2$^7/_8$ in • 70mm	2	Vulnerable

Habitat & Ecology coastal and river delta swamps; larval food plant: *A. macronoura* (member of the birthwort family)
Other Characteristics ABOVE darker ground color; FEMALE similar
Subspecies none described

PARIDES ORELLANA

Size	**Zone**	**Status**
4$^3/_8$ in • 110mm	2	Rare

Habitat & Ecology rain forest
Other Characteristics BELOW reddish pink spot band reduced; FEMALE larger and duller
Subspecies none described

PARIDES PIZARRO

Size	**Zone**	**Status**
3$^7/_8$ in • 95mm	2	Protected

Habitat & Ecology localized populations in rain forest with high endemism in the sub-Andean Amazon basin
Other Characteristics BELOW similar; FEMALE larger wing expanse with more rounded forewings; larger yellow discal patch
Subspecies none described

PARIDES SESOTRIS
Southern Cattle Heart

Size	**Zone**	**Status**
3⁵/₈ in • 90mm	2	Not protected

Habitat & Ecology rain forest; along forest edges; clearings; disturbed secondary forest
Other Characteristics BELOW duller; markings reduced to a pinkish red spot band on hindwing; FEMALE lacks green patches on forewing
Subspecies 4 described

PARIDES STEINBACHI

Size	**Zone**	**Status**
3¹/₈ in • 80mm	2	Protected

Habitat & Ecology Amazonian rain forest
Other Characteristics BELOW paler; markings limited to pinkish red submarginal spot band, and with pink bar on tornus: FEMALE lacks white patch on forewing
Subspecies 2 described

Genus Parnassius

This is a genus of medium-sized, sedentary butterflies, that are normally associated with rocky montane areas, and are even found above the snow line on Mount Everest. These swallowtails are tailless and generally white in color. Their larval food plants are usually members of the *Saxifragaceae* (saxifragaceae family).

PARNASSIUS APOLLO
Apollo

Size	**Zone**	**Status**
3¹/₈ in • 80mm	3	Protected

Habitat & Ecology arctic alpine meadows and grasslands; larval food plant: *Sedum* (roseroot)
Other Characteristics BELOW additional red spot at base; black bar replaced with red; FEMALE darker markings; larger red spots
Subspecies more than 200 described

♀

PARNASSIUS AUTOCRATOR

Size	Zone	Status
5¹/₈ in • 130mm	3	Rare

Habitat & Ecology high montane meadows of Afghanistan and Tajikistan; little known about life history

Other Characteristics BELOW paler; orange markings overlaid with larger white spots; MALE fewer dark markings; reduced orange on hindwing

Subspecies none described

PARNASSIUS EVERSMANNI THOR
Eversmann's Parnassian, Yellow Apollo

Size	Zone	Status
2³/₈ in • 60mm	1•3	Not protected

Habitat & Ecology arctic meadows; tundra
Other Characteristics BELOW more red spots on hindwing; black encircled red bar along inner margin; FEMALE paler; whiter hindwing markings
Subspecies 9 described

PARNASSIUS MNEMOSYNE
Clouded Apollo

Size	Zone	Status
2³/₈ in • 60mm	3	Protected

Habitat & Ecology moist montane meadows
Other Characteristics BELOW similar; FEMALE suffused with gray; darker and more extensive markings
Subspecies more than 100 described

PARNASSIUS PHOEBUS SACERDOS
Small Apollo, Phoebus Parnassian

Size	Zone	Status
3^1/$_8$ in • 80mm	1•3	Vulnerable

Habitat & Ecology open montane meadows; rocky alpine areas; tundra
Other Characteristics BELOW red markings at base of hindwing; extra red spot above tornus; FEMALE hindwing spots completely red
Subspecies more than 40 described

Genus Protographium

This is a small Australian genus with a single species related to *Eurytides* and *Graphium*.

PROTOGRAPHIUM LEOSTHENES
Four-bar, Four-bar Swordtail

Size	Zone	Status
2^5/$_8$ in • 65mm	6	Not protected

Habitat & Ecology rain forest; larval food plant: *Rauwenhoffia leichhardtii* (zig-zag vine)
Other Characteristics ABOVE paler, broader spot bands on hindwing; FEMALE above, lacks red marginal markings on hindwing
Subspecies 2 described

Genus Teinopalpus

This genus, from Southeast Asia, is comprised of two species. The unusual color combinations and markings make these butterflies highly prized by collectors.

TEINOPALPUS AUREUS
Golden Kaiser-I-Hind

Size	Zone	Status
4^3/$_8$ in • 110mm	5	CITES II

Habitat & Ecology variable; males seen in canopy of upland pine rain forests with ridges up to 900 ft (300m); little known about life history; larval food plant: *Magnolia* (magnolia)
Other Characteristics BELOW hindwing markings darker; more vivid golden patch; darker veins; FEMALE unknown
Subspecies 3 described

TEINOPALPUS IMPERIALIS
Kaiser-I-Hind

Size	Zone	Status
5¹/₈ in • 130mm	5	CITES II

Habitat & Ecology canopy of montane rainforest from 300–1,200 ft (100–350m); rapid fliers larval food plant: *Magnolia* (magnolia)
Other Characteristics ABOVE prominent buff median patch overscaled with gray near costa of hindwing, but shading to bright yellow near inner margin; MALE emerald green ground color
Subspecies 4 described

Genus Trogonoptera
The 2 species represented in this genus are associated with *Aristolochia* (birthwort) as the larval food plant.

TROGONOPTERA BROOKIANA ALBESCENS
Rajah Brooke's Birdwing

Size	Zone	Status
7 in • 175mm	5	CITES II

Habitat & Ecology tropical rain forest; males fly in the canopy and congregate at mud puddles in open areas; females are attracted to flowers
Other Characteristics BELOW hindwing completely outlined in white; MALE below, cobalt blue at base of wings; faint indication of bluish green submarginal spot band
Subspecies 5 described

TROGONOPTERA TROJANA

Size	Zone	Status
8 in • 200mm	5	CITES II

Habitat & Ecology rain forest canopies
Other Characteristics BELOW forewing markings reduced; a few iridescent aquamarine postmedian spots; FEMALE veins outlined in cream at apex
Subspecies none described

Genus Troides

Comprising 20 Southeast Asian species, this genus is characterized by black forewings and some yellow to golden markings on the hindwing.

TROIDES DOHERTYI
Talaud Black Birdwing

Size	**Zone**	**Status**
6⅝ in • 165mm	5	CITES II

Habitat & Ecology low and coastal rain forest
Other Characteristics BELOW similar; FEMALE dull brown forewing; paler between veins; darker hindwing; reduced dull yellow median patch
Subspecies 5 described

TROIDES AEACUS
Golden Birdwing

Size	**Zone**	**Status**
6 in • 150mm	5	CITES II

Habitat & Ecology rain forest; little known about ecology and life history
Other Characteristics BELOW forewing veins along margin heavily outlined in lustrous grayish white; FEMALE diffuse orange patches on hindwing
Subspecies 5 described

TROIDES HELENA CERBERUS
Common Birdwing

Size	**Zone**	**Status**
7 in • 175mm	5•6	CITES II

Habitat & Ecology variable, from tropical rain forest to marginal secondary forests
Other Characteristics BELOW forewing veins outlined in lustrous gray; FEMALE veins outlined in dusky cream
Subspecies 7 described

TROIDES HYPOLITUS

Size	Zone	Status
8 in • 200mm	5•6	CITES II

Habitat & Ecology rain forest
Other Characteristics BELOW similar;
FEMALE forewing veins outined in dusky white;
outline of veins expanded below
Subspecies 3 described

TROIDES PRATTORUM
Buru Opalescent Birdwing

Size	Zone	Status
8 in • 200mm	6	CITES II

Habitat & Ecology rain forest (Buru Island)
Other Characteristics BELOW lacks brown scales
above tornus; FEMALE black ground color;
forewing veins outlined in cream; paler yellow
hindwing gold patch
Subspecies none described

Genus Zerynthia

This small Eurasian genus comprises
2 species and is associated with the
noxious *Aristolochia* (birthwort) as larvae.

ZERYNTHIA RUMINA
Spanish Festoon

Size	Zone	Status
1⁷/₈ in • 45mm	3	Not protected

Habitat & Ecology montane meadow; rocky
slopes
Other Characteristics BELOW enlarged red
markings; MALE cream ground color; red
markings reduced
Subspecies 14 described

♀

CHAPTER 2

Pieridae

The Pieridae, commonly called the whites and sulfurs, are distinguished readily by their generally bright white and yellow coloration. However, members of this family, that range widely throughout the tropics worldwide, are highly variable in coloration and are often involved in mimetic associations with representatives of other families. The sexes are generally dimorphic. Some species, such as the Large White and the Small White, are crop pests, while others use a wide variety of larval food plants. A number of species are highly migratory, and these tendencies enable them to expand their ranges markedly. Most pierids are gregarious, are avid flower visitors, and are often observed in large numbers around water, especially at mud puddles.

Genus Anteos
The 2 large, angular sulfur butterflies that make up this genus are generally observed in open habitats in Central and South America and the southern United States. They are very strong fliers, and often migratory.

ANTEOS CLORINDE NIVIFERA
Ghost Brimstone, White-angled Sulfur

Size	Zone	Status
3⅝ in • 90mm	1•2	Not protected

Habitat & Ecology open scrubby woodlands; often seen at mud puddles and streams
Other Characteristics BELOW pale, mottled to simulate a leaf; FEMALE pale green ground color; eyespots encircled with gold
Subspecies 12 described

Genus Anthocharis
These are the "orange tips," so named for the bright colored patches below the apex of the forewing. These butterflies are restricted to the Northern Hemisphere and are found in open woodlands and montane areas.

ANTHOCHARIS CARDAMINES
Orange Tip

Size	Zone	Status
2 in • 50mm	3	Not protected

Habitat & Ecology open country; hedges
Other Characteristics BELOW white ground color; pale orange apex on forewing; olive green mottling on hindwing; FEMALE lacks the orange tips on the forewings; below, black and white; lacks mottling
Subspecies more than 20 described

ANTHOCHARIS MIDEA
Falcate Orange Tip

Size	Zone	Status
1¾ in • 40mm	1	Not protected

Habitat & Ecology woodland; roadside hedges; disturbed areas; widely found
Other Characteristics BELOW olive green mottling on hindwing; FEMALE lacks orange tips
Subspecies 2 described

Genus Aphrissa

The 10 species of this genus are widespread in the Central and South American tropics. The larvae are polymorphic and associated with *Bignoniaceae* (trumpet creeper family) and *Fabaceae* (legume/pea family).

APHRISSA STATIRA
Statira, Migrant Sulfur, Yellow Migrant

Size	Zone	Status
2⁷/₈ in • 70mm	1•2	Not protected

Habitat & Ecology open areas; sparse woodland; larval food plant: *Cassia* (cinnamon)
Other Characteristics BELOW ground color yellow; white along forewing inner margin; fine pink markings; FEMALE uniformly yellow with a thin black border on the forewing
Subspecies more than 4 described

Genus Appias

This is a distinctive genus of butterflies with strongly curved wings, especially in the males, which greatly enhances their flight.

APPIAS CELESTINA
Common Migrant

Size	Zone	Status
2⁷/₈ in • 70mm	5•6	Not protected

Habitat & Ecology rain-forest edges
Other Characteristics BELOW faint indication of markings above; FEMALE 2 forms: yellow or white; broader black margins; black at base; white or yellow apical band
Subspecies 8–10 described

Genus Aporia

This is large genus with more than 25 species distributed throughout Europe and Asia.

APORIA CRATAEGI
Black-veined White

Size	Zone	Status
2⁷/₈ in • 70mm	3	Not protected

Habitat & Ecology open country; can be an orchard pest; larval food plants: *Spiraea* (member of the rose family), *Prunus* (plum family), and *Crataegus laevigata* (hawthorn)
Other Characteristics BELOW darker veins; black scales on hindwing; FEMALE grayish white; larger by up to 4 in (100mm)
Subspecies more than 20 described

APPIAS DRUSILLA
Tropical White, Florida White

Size	Zone	Status
2³/₈ in • 60mm	1•2	Not protected

Habitat & Ecology open woodland; forest edges; avid flower visitor; larval food plants: *Drypetes alba* (cafeillo), *Drypetes lateriflora* (guiana plum)
Other Characteristics BELOW unmarked white; fine black line along costa; FEMALE dusky grayish; often a yellow flush at forewing base
Subspecies 8 described

APPIAS PHAOLA
Congo White

Size	Zone	Status
2³/₈ in • 60mm	4	Not protected

Habitat & Ecology open woodlands; along paths
Other Characteristics BELOW forewing yellow at base; black marginal spots at vein ends; FEMALE above, off-white with a pale yellow submarginal spot band
Subspecies 3 described

APPIAS NERO
Orange Albatross

Size	Zone	Status
2⁵/₈ in • 65mm	5•6	Not protected

Habitat & Ecology forest; females in the canopy
Other Characteristics ABOVE unmarked pale orange; FEMALE above, dull orange with a darker margin; cell end patch on forewing
Subspecies more than 20 described

Genus Archonias

This genus of medium-sized butterflies from Central and South America is characterized by the flattened antennae and slightly scalloped hindwing margin.

▲
♀

ARCHONIAS ERYCINIA

Size	Zone	Status
2 5/8 in • 65mm	2	Not protected

Habitat & Ecology rain forest; males often observed patrolling and perched along paths in the understorey; not found north of Mexico; life history incomplete; only larval food plant on record: *Loranthaceae* (mistletoe family)
Other Characteristics ABOVE black with yellow bar on forewing; MALE similar
Subspecies 3 described

Genus Ascia

A genus of whites that live in open areas and range widely in North and South America; they are strong fliers and highly migratory. The larvae are normally associated with the *Cruciferae* (mustard family).

▲
♀

ASCIA MONUSTE
Great Southern White

Size	Zone	Status
2 5/8 in • 65mm	1•2	Not protected

Habitat & Ecology disturbed habitats; gardens
Other Characteristics ABOVE white ground color; black along costa, at apex, and along lateral margins; MALE similar
Subspecies none described

Genus Belenois

This African genus of medium-sized butterflies comprises 19 species, some of which are migratory.

BELENOIS CREONA REVERINA
African Common White

Size	Zone	Status
2 3/8 in • 60mm	4	Not protected

Habitat & Ecology open coastal areas; larval food plant: *Capparis* (caper)
Other Characteristics BELOW hindwing ground color pale yellow; darkened veins; FEMALE may be white or creamy yellow with more extensive dark markings; below, veins more heavily outlined in black
Subspecies more than 7 described

Genus Catasticta

This is a large genus of butterflies with high species diversity that ranges from Mexico, through Central and South America. The larvae are gregarious and associated with the *Loranthaceae* (mistletoe family). The pupae resemble bird droppings.

CATASTICTA URICOECHEAE

Size	Zone	Status
2 in • 50mm	2	Rare

Habitat & Ecology forest
Other Characteristics BELOW gray forewing; end cell bar white; double postdiscal spot bands with inner pointed spots; FEMALE above, more red
Subspecies none described

BELENOIS RAFFRAYI
Raffray's White

Size	Zone	Status
2¹/₈ in • 55mm	4	Not protected

Habitat & Ecology forest
Other Characteristics BELOW yellow suffusion at base of hindwing costa; FEMALE similar
Subspecies 3 described

Genus Catopsilia

As the common name "mottled emigrant" implies, these butterflies are migratory and widely distributed from the Canary Islands to Australia, including Africa and southern Europe.

CATOPSILIA FLORELLA
African Immigrant

Size	Zone	Status
2⁷/₈ in • 70mm	4•5	Not protected

Habitat & Ecology variable, as is very migratory
Other Characteristics BELOW forewing unmarked white; FEMALE below, orange yellow, white along tornus and inner margin of hindwing; forewing cell end and postmedian band brown; hindwing with brown postdiscal band and two additional spots near cell end
Subspecies none described

CATOPSILIA SCYLLA
Yellow Migrant, Orange Emigrant

Size	Zone	Status
2⁵/₈ in • 65mm	5•6	Not protected

Habitat & Ecology forest clearings
Other Characteristics BELOW golden yellow ground color; silver cell end spot encircled in rust brown and an irregular subapical brown band on forewing; FEMALE duller; broader dark margins on both wings; submarginal dark band on the forewing
Subspecies 10 described

Genus Cepora

The 20 species of this genus are slow-flying pierids that range from India and Sri Lanka eastward to Norfolk Island and the Fiji Islands. The larval food plants include the *Capparidaceae* (caper family).

Genus Colias

This is a very large and successful genus comprising 75 species and found in all biogeographic regions, except the Australian region.

CEPORA JUDITH

Size	Zone	Status
2³/₈ in • 60mm	5	Not protected

Habitat & Ecology lowland forests; populations often occur in isolated areas
Other Characteristics BELOW markings intensified; FEMALE duller in coloration
Subspecies 15–20 described

COLIAS CROCEUS
Clouded Yellow

Size	Zone	Status
2¹/₈ in • 55mm	3•5	Not protected

Habitat & Ecology open country; often observed visiting waysides and flowers
Other Characteristics BELOW red rimmed silver spot at end cell; FEMALE duller; yellow spots on outer black wing margin; markings intensified
Subspecies none described

COLIAS DIMERA
Small Andean Sulfur

Size	Zone	Status
1⁷/₈ in • 45mm	2	Not protected

Habitat & Ecology montane meadows
Other Characteristics BELOW unmarked orange with pink fringes; reddish pink streaks on hindwing; FEMALE orange, white, or greenish ground color; markings intensified
Subspecies none described

COLIAS ELECTO
*African Clouded Yellow,
Lucerne Butterfly*

Size	Zone	Status
2 in • 50mm	4	Not protected

Habitat & Ecology open fields; avid flower visitor
Other Characteristics BELOW silver cell end encircled in pink brown; a pink brown subapical spot; 4–5 similar postdiscal spots on pinkish hindwing fringe; FEMALE variable ground color; intensified markings
Subspecies 8 described

COLIAS NASTES
*Arctic Green Sulfur,
Pale Arctic Clouded Yellow*

Size	Zone	Status
1⁷/₈ in • 45mm	1•3	Not protected

Habitat & Ecology arctic tundra; montane areas
Other Characteristics BELOW silver spot at end cell of hindwing; black apical spot and red fringe; FEMALE markings intensified
Subspecies 6–8 described

COLIAS PALAENO
Palaeno Sulfur, Chippewa Sulfur

Size	Zone	Status
2 in • 50mm	1•3	Not protected

Habitat & Ecology tundra; taiga; boreal habitats
Other Characteristics BELOW faint black cell
spot and pink along margins and at apex of
forewing; hindwing peppered with gray scales;
FEMALE color variable; markings intensified
Subspecies 11 described

COLIAS PHILODICE
Clouded Sulfur

Size	Zone	Status
2 in • 50mm	1•2	Not protected

Habitat & Ecology open fields and country; larval
food plant: *Melilotus officinalis* (clover)
Other Characteristics BELOW 3 postdiscal spots
on forewing; hindwing overscaled by gray with
double cell end spot; FEMALE variable ground
color; markings intensified
Subspecies 4 described

Genus Colotis

This genus of 50 species is found
primarily in Africa and Asia, with a
few species in southern Europe.
Populations are large wherever they
occur, and often exhibit astonishing
variation. Larval food plants generally
include members of the *Capparidaceae*
(caper family).

COLOTIS ANTEVIPPE
Red Tip, Large Orange Tip

Size	Zone	Status
1³/₄ in • 40mm	4	Not protected

Habitat & Ecology lowland savannah
Other Characteristics BELOW faint orange at tip
of forewing and faint indication of cell end spot;
yellow hindwing overscaled with gray;
FEMALE below, forewing orange at apex; black
markings below cell; cell end spot; tan
hindwing, white outside cell, red-brown V-
shaped line around end of cell; veins darkened
Subspecies 2 described

COLOTIS AURIGINEUS
Veined Gold, Double-banded Orange

Size	Zone	Status
2 in • 50mm	4	Not protected

Habitat & Ecology savannah up to 1,000 ft (310m)
Other Characteristics BELOW ends of veins darkened; FEMALE duller in color; markings expanded
Subspecies 2 described

COLOTIS CELIMENE
Lilac Tip, Magenta Tip

Size	Zone	Status
1⁷/₈ in • 45mm	4	Not protected

Habitat & Ecology dry savannah; subdesert
Other Characteristics BELOW yellow at base of wings; faint black cell end spot with subapical purple patch; dark bands on hindwing; veins darkened; FEMALE lacks the magenta tip
Subspecies 4 described

COLOTIS CALAIS
Banded Gold Tip

Size	Zone	Status
2¹/₈ in • 55mm	4	Not protected

Habitat & Ecology dry savannah; subdesert; fast flier; occasional flower visitor
Other Characteristics BELOW cream forewing apex and hindwing costa; black points at the end of veins on both wings; FEMALE above, cream ground color
Subspecies 4 described

COLOTIS HALIMEDE
Yellow Patch White

Size	Zone	Status
2¹/₈ in • 55mm	4	Not protected

Habitat & Ecology very dry savannah conditions; occasionally observed at midday congregated in deep shade
Other Characteristics BELOW white ground color; unmarked; FEMALE frequently with less yellow on wings
Subspecies 4 described

COLOTIS IONE
Bushveld Purple Tip, Purple Tip

Size	Zone	Status
2⁷/₈ in • 70mm	4	Not protected

Habitat & Ecology savannah; very common in its range
Other Characteristics BELOW violet at apex faintly visible; FEMALE ground color varies from grayish white to yellow; gray or black at base with gray black markings along margins of both wings; forewing apex variable with white; orange; or pale yellow submarginal spot bands
Subspecies 2 described

Genus Delias

This is a genus of medium-sized butterflies comprising over 70 species. They have brightly colored under hindwings with remarkable patterns. Larvae are gregarious, associated with the *Loranthaceae* (mistletoe family), and very distasteful to birds.

DELIAS ARUNA

Size	Zone	Status
3 in • 75mm	6	Not protected

Habitat & Ecology primary rain forest
Other Characteristics BELOW black ground color; narrow yellow cell end bar on forewing; red patch near basal costa on hindwing; FEMALE 2 color forms: (1) duller orange at base of wings; broadly black toward the tips of both wings; (2) above, dusky yellow to cream with black margins on both wings; white spot at forewing end cell
Subspecies 7 described

DELIAS BAGOE

Size 3 1/8 in • 80mm **Zone** 6 **Status** Not protected

Habitat & Ecology open areas
Other Characteristics ABOVE cream ground color; large subapical yellow patch surrounded in black; hindwing cream with a black border; FEMALE above, fawn-colored surface enclosed finely in black; markings more extensive
Subspecies 4 described

DELIAS HARPALYCE
Imperial White

Size 3 1/8 in • 80mm **Zone** 6 **Status** Not protected

Habitat & Ecology found near their larval food plant: *Loranthaceae* (mistletoe family); common around the Great Dividing Range in Australia
Other Characteristics ABOVE darker; lighter cell on forewing with a submarginal spot band; hindwing grayish with a broad black border; MALE white with narrow black margins, and a white submarginal spot band
Subspecies none described

♀

DELIAS HENNINGIA

Size 3 1/8 in • 80mm **Zone** 5 **Status** Not protected

Habitat & Ecology Philippine rain forest
Other Characteristics ABOVE gray blue in, and just outside, cell; FEMALE variable color patterns; larger; darker; above, blackish with faint powder blue diagonal band on forewing; a prominent yellow patch in cell of hindwing
Subspecies 5 described

Genus Dercas

This is a small genus of 4 East Asian species.

♀

DERCAS LYCORIAS

Size	Zone	Status
2³/₈ in • 60mm	5	Not protected

Habitat & Ecology dense forest and forest edges up to 4,500 ft (1,500m)
Other Characteristics BELOW markings brown; MALE forewing more curved near apex
Subspecies 4 described

Genus Dismorphia

This genus of more than 23 species mimics *heliconiines* and *ithomiines* and may also be somewhat distasteful. Most species are associated with *Mimoseae* (member of the legume family) as larvae.

DISMORPHIA CORDILLERA

Size	Zone	Status
3 in • 75mm	2	Not protected

Habitat & Ecology rain forest
Other Characteristics BELOW similar; FEMALE lacks forewing angular margin, and polished androconial area on the hindwing costa
Subspecies 7 described

DISMORPHIA SPIO
Haitian Mimic

Size	Zone	Status
2⁷/₈ in • 70mm	2	Not protected

Habitat & Ecology dense forest
Other Characteristics 3 different forms occur in both sexes; BELOW hindwing mottled brown and ochreous, frosted with white; FEMALE similar
Subspecies none described

Genus Dixeia
This African genus of 8 species is also found in Madagascar. They are typically white with some shading of yellow. Larval food plants include members of the *Capparidaceae* (caper family).

DIXEIA DOXO COSTATA
Black-veined White, African Small White

Size	Zone	Status
2 in • 50mm	4	Not protected

Habitat & Ecology open country; gardens; flower visitor; larval food plant: *Capparis* (caper)
Other Characteristics ABOVE blackened white; FEMALE expanded broad apex and submargin; enlarged black postdiscal spot on forewing; black cell end spot on both wings; veins darkened
Subspecies 6 described

DIXEIA SPILLERI
Spiller's Sulfur Yellow

Size	Zone	Status
1⁷⁄₈ in • 45mm	4	Not protected

Habitat & Ecology common savannah species
Other Characteristics BELOW similar; FEMALE often duller; varies from yellow to cream to ochreous yellow
Subspecies none described

Genus Elphinstonia
This is a small genus of butterflies comprising 4 species and related to *Euchloe*.

ELPHINSTONIA CHARLONIA
Greenish-black Tip

Size	Zone	Status
1³⁄₈ in • 35mm	3	Not protected

Habitat & Ecology open flowery meadows
Other Characteristics BELOW hindwing greenish gray with white cell end spot; white spot mid-costa; FEMALE larger; forewings less acute
Subspecies 9 described

Genus Eroessa

This genus is represented by a single, rare species and is endemic to the high mountains of Chile.

EROESSA CHILIENSIS

Size	Zone	Status
1³/₈ in • 35mm	2	Not protected

Habitat & Ecology montane meadow; rocky terrain
Other Characteristics BELOW additional white spots from apex to mid-submargin; hindwing white with darkened veins and chevrons near ends of cell; FEMALE yellow tip on forewing apex
Subspecies none described

Genus Eronia

A genus of African pierids that are widely distributed as far as Saudi Arabia, these are found in a variety of habitats.

ERONIA CLEODORA
Vine-leaf Vagrant

Size	Zone	Status
2⁷/₈ in • 70mm	4	Not protected

Habitat & Ecology open country to subdesert, woodlands, or on the edge of the bush in South Africa; settles on flowers close to the ground
Other Characteristics ABOVE 2 silver spots near base of hindwing; FEMALE ground color variable from white, to cream, to yellow
Subspecies none described

Genus Euchloe

Found in Eurasia, with a few represented in North America, these whites are characterized by green and white marbling on the under hindwing.

EUCHLOE AUSONIA
Dappled White

Size	Zone	Status
2 in • 50mm	1•3	Not protected

Habitat open montane meadows and waysides; larval food plant: *Cruciferae* (mustard family)
Other Characteristics *E. ausonides* may be conspecific with this species; BELOW black cell end spot and greenish mottling at apex of forewing; hindwing white with net patterns of green throughout; FEMALE duller and larger
Subspecies more than 20 described

EUCHLOE BELEMIA
Green-striped White

Size	Zone	Status
1⁷/₈ in • 45mm	3•4	Not protected

Habitat & Ecology open country; montane areas
Other Characteristics ABOVE gray at forewing apex and along margins; white submarginal spots; FEMALE ground color varies
Subspecies 8 described

Genus Eurema

This is a very large genus of small yellow, orange, and/or white butterflies that range worldwide. They are often common where they occur.

EUREMA DAIRA
Barred Yellow

Size	Zone	Status
1⁷/₈ in • 45mm	1•2	Not protected

Habitat & Ecology open country; a frequent visitor to wild flowers
Other Characteristics polymorphic; BELOW generally white with faint indication of markings above; marked seasonal variation and dimorphism; FEMALE lacks gray forewing bar
Subspecies 4 described

EUREMA MEXICANA
Mexican Yellow

Size	Zone	Status
2 in • 50mm	1•2	Not protected

Habitat & Ecology xeric open areas
Other Characteristics BELOW white forewing with yellow at base and along margin; hindwing light yellow with a few brown spots; FEMALE lacks drooping dog face on wings
Subspecies 2 described

EUREMA PROTERPIA
Sleepy Orange, Jamaican Orange,
Tailed Orange

Size	**Zone**	**Status**
2 in • 50mm	1•2	Not protected

Habitat & Ecology desert scrub
Other Characteristics originally described as
E. gundlachiana; tailed morphs in both sexes
BELOW forewing unmarked; slightly darkened on
costal margin and at apex; hindwing paler
orange; can be mottled; FEMALE duller; below,
lacks patterning
Subspecies 1 described

EUREMA SENEGALENSIS
Common Grass Yellow,
Forest Grass Yellow

Size	**Zone**	**Status**
1⁷/₈ in • 45mm	4•5	Not protected

Habitat & Ecology a common migrant found in
open country; savannah; scrub; gardens;
disturbed areas
Other Characteristics r yellow wings with wavy-
edged brown black apical mark; FEMALE larger
Subspecies none described

Genus Ganyra

This is a genus of very large, neotropical,
white butterflies closely related to *Ascia*.

GANYRA JOSEPHINA
Giant White

Size	**Zone**	**Status**
2⁷/₈ in • 70mm	2	Not protected

Habitat Open woodlands; desert scrub
Other Characteristics BELOW white markings;
MALE paler buff; below, patterns evident from
above
Subspecies 5 described

Genus Gonepteyrx

These are the familiar Brimstone butterflies. There are 6 species recognized in North Africa and Eurasia.

GONEPTEYRX RHAMNI
Brimstone

Size	Zone	Status
2³/₈ in • 60mm	3•5	Not protected

Habitat & Ecology open woodland; disturbed areas; larval food plant: *Rhamnus cathartica* (buckthorn)
Other Characteristics BELOW yellowish green; unmarked except for red brown cell end spots on both wings; FEMALE above, white; below, greenish gray
Subspecies 12 described

Genus Hebomia

This is a small genus of three of the largest pierids in the world. They are very strong fliers.

HEBOMIA GLAUCIPPE
Great Orange Tip

Size	Zone	Status
4 in • 100mm	5•6	Not protected

Habitat & Ecology streams; woodland edges
Other Characteristics BELOW forewing apex tan, peppered with black; hindwing tan peppered with black spots with a black central band from base, across cell to margin; FEMALE cream ground color on hindwing; reduced orange on forewings; expanded black margins on both wings; a spot band of black on hindwing
Subspecies 3 described

HEBOMIA LEUCIPPE DETANII

Size	Zone	Status
4 in • 100mm	5•6	Not protected

Habitat & Ecology streams in Indonesia
Other Characteristics BELOW grayish tan costal and postdiscal spots, with tan median stripe from base, across cell to margin, on hindwing; FEMALE duller
Subspecies 3–4 described

Genus Hesperocharis

This neotropical genus of 15 species is infrequently encountered. Consequently, little is known about its biology and life history.

HESPEROCHARIS GRAPHITES

Size
2⁵/₈ in • 65mm

Zone
2

Status
Not protected

Habitat & Ecology open patches of rain forest
Other Characteristics ABOVE clear yellow; hindwing markings darkened at end veins; FEMALE larger and duller; with more rounded wings
Subspecies 2 described

Genus Itaballia

This is a genus comprising 7 species in which most females are both dimorphic and highly mimetic.

ITABALLIA VIARDI

Size
2⁵/₈ in • 65mm

Zone
2

Status
Not protected

Habitat & Ecology tropical scrub; xeric forest
Other Characteristics BELOW apex and hindwing paler and overscaled with white and gray; faint yellow submarginal spot on both wings; MALE dimorphic; above, white with black forewing apex, few white spots, and black triangular spot end cell
Subspecies 2 described

Genus Ixias

This genus is represented by 12 species and is closely related to the African genus *Colotis*.

IXIAS PYRENE SESIA
Yellow Orange Tip

Size
2 in • 50mm

Zone
5

Status
Not protected

Habitat & Ecology open woodlands near water
Other Characteristics BELOW similar; FEMALE highly variable; may be quite dark or very pale with rounder wing shape; strongly marked
Subspecies 15 described

IXIAS REINWARDTII PAGENSTECHERI

Size	**Zone**	**Status**
2¹/₈ in • 55mm	5•6	Not protected

Habitat & Ecology closely associated with forest
Other Characteristics BELOW pale yellow ground color; large brown cell spot on forewing; small spot on hindwing cell; brown postmedian bands around cells of both wings; FEMALE paler
Subspecies 6 described

Genus Klotsius

Originally associated with *Anteos*, *Klotsius* was recently described on the basis of the wing shape and other structural differences.

KLOTSIUS MENIPPE
Mammoth Sulfur

Size	**Zone**	**Status**
5¹/₈ in • 130mm	2	Not protected

Habitat & Ecology fields; trails; jungle paths
Other Characteristics BELOW paler ground color; brown cell end spots on both wings; larger on forewing; FEMALE duller; brown speckles below
Subspecies none described

Genus Leptosia

Members of this genus, endemic to parts of Africa and Asia, are white and very delicate in appearance.

LEPTOSIA ALCESTA
African Wood White

Size	**Zone**	**Status**
1⁷/₈ in • 45mm	4•5	Not protected

Habitat & Ecology open woodlands; jungle trails
Other Characteristics BELOW off-white; black postdiscal forewing; grayish hindwing with a thin band across end cell; FEMALE larger
Subspecies 5 described

Genus Melete

This neotropical genus of yellow and white butterflies is characterized by a black bar at the end of the ventral forewing cell.

MELETE LYCIMNIA

Size	Zone	Status
2⁷/₈ in • 70mm	2	Not protected

Habitat & Ecology open forest
Other Characteristics ABOVE cream ground color; black markings repeated above; faint forewing cell spot; FEMALE broader black margin
Subspecies 7 described

Genus Mylothris

This African genus, comprising 50 species, is noted for its many different combinations of white and yellow. They are slow fliers, which could be related to their distasteful larval food plant *Loranthaceae* (mistletoe family).

♀

♀

MYLOTHRIS AGATHINA
Common Dotted White, Common Dusted Border

Size	Zone	Status
2¹/₈ in • 55mm	4	Not protected

Habitat & Ecology rain forest; often migratory
Other Characteristics BELOW pinkish orange ground color; veins darkened; MALE orange more intense; median forewing white
Subspecies 4 described

MYLOTHRIS RHODOPE
Rhodope, Tropical Dotted Border

Size	Zone	Status
2⁵/₈ in • 65mm	4	Not protected

Habitat & Ecology tropical rain forest; flies high in the canopy, perhaps to entice predators
Other Characteristics BELOW orange forewing; white hindwing; small black spots at end veins; MALE below, white with yellow at base of wings; veins indicated by black points on wing margins
Subspecies 8 described

Genus Neophasia

As their common names suggest the two species of this genus are associated with *Pinus* (pine) as the larval food plant and found in western North America and the montane northwestern part of Mexico.

NEOPHASIA MENAPIA
Pine White

Size	Zone	Status
2 in • 50mm	1	Not protected

Habitat & Ecology common in pine woodlands, often flying high in these forests
Other Characteristics BELOW hindwing darker with black submarginal band; FEMALE hindwing veins outlined in pink and black
Subspecies 2 described

NEOPHASIA TERLOOII
Chiricahua Pine White, Mexican Pine White

Size	Zone	Status
2³⁄₈ in • 60mm	1•2	Not protected

Habitat & Ecology pine woods; males pursue females agressively
Other Characteristics ABOVE veins darkened on both wings; MALE white, with a very distinctive black forewing cell; below, white with black; veins darkened; dimorphic
Subspecies none described

Genus Nepheronia

These butterflies are relatively large and characterized by combinations of white and yellow with black margins. Primarily from Africa with one species found in India, they generally migrants and are strong fliers.

NEPHERONIA THALASSINA
Cambridge Vagrant

Size	Zone	Status
2³⁄₈ in • 60mm	4	Not protected

Habitat & Ecology woodland edges; migratory
Other Characteristics sexes with prominent gloss or sheen; BELOW ground color glossy yellowish white; faint indication of markings; FEMALE variable; usually yellow orange apex with three brownish subapical/submarginal spots; hindwing glossy white; darker yellow along margin
Subspecies none described

Genus Pareronia

The females of these 10 species from Asia and Africa are highly polymorphic.

PARERONIA TRITAEA

Size
4 in • 100mm

Zone
5

Status
Not protected

Habitat & Ecology rain forest on Sulawesi
Other Characteristics BELOW pale bluish green; all veins have a dusky or black outline; FEMALE brownish black; bluish spots in cells; blue submarginal spot band
Subspecies 7 described

Genus Pereute

Endemic to Central and South America, this genus comprises 10 species and is characterized by the black ground color and contrasting forewing transverse bands.

PEREUTE LEUCODROSIME
Red-banded Pereute

Size
3 in • 75mm

Zone
2

Status
Not protected

Habitat & Ecology rain forest, especially in Colombia and Ecuador
Other Characteristics BELOW dark brown ground color; red transverse band; red spots at base of each wing; veins darkened; FEMALE similar
Subspecies 3 described

Genus Phoebis

Restricted to the Western Hemisphere, this genus comprises 15 species of structurally distinct, large yellow butterflies with a variety of markings.

PHOEBIS AGARITHE
Large Orange Sulfur

Size
$2^7/8$ in• 70mm

Zone
1•2

Status
Not protected

Habitat & Ecology disturbed areas; attracted to flowers; often migratory
Other Characteristics BELOW light brown markings; cell end spots and postmedian bands on both wings; FEMALE veins darkened at margin; larger markings enhanced with fine brown speckling
Subspecies 4 described

PHOEBIS AVELLANEDA
Red-splashed Sulfur

Size	Zone	Status
3⁵/₈ in • 90mm	2	Not protected

Habitat & Ecology montane woodlands; rapid flier; does not frequent flowers; not migratory
Other Characteristics BELOW silver cell end spots ringed; brown apical and postmedian band on forewing; brown median band on hindwing; veins darkened; FEMALE dimorphic; heavier markings; above, brick red
Subspecies none described

PHOEBIS RURINA

Size	Zone	Status
3¹/₈ in • 80mm	2	Not protected

Habitat & Ecology open country; larval food plant: *Cassia fruticosa* (member of legume family)
Other Characteristics BELOW prominent red brown cell end spots; FEMALE dimorphic; heavier markings
Subspecies 2 described

Genus Pieris

Generally known as the cabbage butterflies, members of this genus are pests on crops, and are quite distinctive in their appearance.

PIERIS BRASSICAE
Large White, Cabbage White

Size	Zone	Status
2⁵/₈ in • 65mm	2•3•4•5	Not protected

Habitat & Ecology open disturbed areas; gardens; highly migratory
Other Characteristics BELOW 2 black postdiscal spots; reduced black spot near hindwing apex; FEMALE dimorphic; heavier markings
Subspecies 10 described

PIERIS RAPAE
Sharp-veined White, Small White

Size	Zone	Status
2 1/8 in • 55mm	1•3•5	Not protected

Habitat & Ecology open country; meadows; gardens
Other Characteristics BELOW pale yellow suffusion at apex of forewing; hindwing veins darkened; FEMALE with larger, darker spots; below, less dusky on hindwing
Subspecies more than 20 described

PIERIS VIRGINIENSIS
West Virginia White

Size	Zone	Status
1 7/8 in • 45mm	1	Not protected

Habitat & Ecology mesic open woodlands; larval food plant: *Lathraea* (toothwort)
Other Characteristics BELOW hindwing veins outlined in gray; FEMALE similar
Subspecies none described

Genus Pinacopteryx
This small, distinctive, African genus is associated with the *Capparidaceae* (caper family) as larvae. There are seasonally variable forms, those observed during the dry season being considerably lighter.

PINACOPTERYX ERIPHIA
Zebra White

Size	Zone	Status
2 5/8 in • 65mm	4	Not protected

Habitat & Ecology open scrub; avid flower visitor
Other Characteristics BELOW yellow more prominent; red-brown markings; FEMALE similar
Subspecies 4 described

Genus Pontia

These small, white, migratory butterflies with mottled green undersurface range widely throughout North America, Europe, Asia, and Africa. Larvae are associated with various members of the *Cruciferae* (mustard family).

PONTIA HELICE
Mustard White, Meadow White

Size	Zone	Status
2 in • 50mm	4	Not protected

Habitat & Ecology montane flowering meadows; a pest in gardens
Other Characteristics BELOW apex veins outlined in brown with red; hindwing veins darkened; FEMALE similar
Subspecies 2 described

PONTIA CALLIDICE
Peak White

Size	Zone	Status
2 in • 50mm	1•3•5	Not protected

Habitat & Ecology open meadows, especially montane areas; attracted to flowers
Other Characteristics BELOW hindwing cell; veins along costa and submargin outlined in olive green; FEMALE darker markings, especially on forewing apex; gray green markings heavier and more extensive
Subspecies 9 described

PONTIA OCCIDENTALIS
Western White

Size	Zone	Status
1⁷/₈ in • 45mm	1	Not protected

Habitat & Ecology meadows; open woods
Other Characteristics BELOW hindwing cell and veins outlined in olive green; with prominent postdiscal chevron spot band; FEMALE slightly darker markings on forewing
Subspecies 2 described

Genus Prestonia

This is a rare monotypic genus identified with a specialized habitat in southwestern Mexico.

PRESTONIA CLARKI

Size	Zone	Status
2⁷/₈ in • 70mm	2	Not protected

Habitat & Ecology woods and forest edges
Other Characteristics sexes with pale, pinkish spotting in, and around the under hindwing cell; BELOW reddish pink discal spots; FEMALE paler; below, heavier markings
Subspecies none described

Genus Prioneris

This is a genus of large butterflies from India to Malaysia that are normally associated with the *Capparidaceae* (caper family) as larvae.

PRIONERIS CLEMANTHE
Redspot Sawtooth

Size	Zone	Status
3¹/₈ in • 80mm	5	Not protected

Habitat & Ecology forest canopy
Other Characteristics ABOVE similar; FEMALE above, grayer, especially on the forewing; broader, slightly yellowish hindwing margin
Subspecies 2 described

Genus Zerene

Widely distributed throughout North and South America, this genus is characterized by the prominent dog face on the forewing.

ZERENE CESONIA
Dog-face Butterfly

Size	Zone	Status
2⁷/₈ in • 70mm	1•2	Not protected

Habitat & Ecology open meadows and woodland areas; several from the Andes
Other Characteristics BELOW silver cell end spots on both wings; FEMALE below, pinkish flush
Subspecies 6-8 described

CHAPTER 3

Nymphalidae

These are the brush-footed butterflies, this refers to the reduced pair of first legs or forelegs that are covered in dense hairs or setae. Although most butterfly families have three functional pairs of walking legs, the Nymphalidae appear to have only two when at rest on flowers or leaves. Otherwise, it is difficult to generalize about the characters that define this family. This is an amalgamation of several groups that have previously been considered as separate familes, and includes the satyrids, danaids, ithomiids, heliconiids, limenitiids, charaxids, and apaturids. The snout butterflies or Libytheidae are included here, but some authors consider them as a separate family.

Genus Acraea

This primarily African genus, with a few representatives in Asia, comprises more than 150 species that sequester toxic compounds and are poisonous to predators. They are associated with members of the *Passifloraceae* (violales) as larvae.

ACRAEA ACRITA
Fiery Acraea

Size	Zone	Status
2⅝ in • 65mm	4	Not protected

Habitat & Ecology open savannah
Other Characteristics BELOW pale yellow flush in cell; yellow spot band on hindwing; FEMALE similar
Subspecies none described

ACRAEA ANDROMACHE
Glasswing, Glass Wing, The Small Greasy

Size	Zone	Status
2⅜ in • 60mm	6	Not protected

Habitat & Ecology scrub; open low woodlands; forests; areas surrounding streams of northern Australia, Indonesia, and New Guinea; larval food plant: *Passiflora* (passion flower)
Other Characteristics BELOW paler; larger yellow spots on hindwing margin; FEMALE similar
Subspecies 3 described

Genus Adelpha

Composed of more than 80 species with similar wing patterns, this genus can be difficult to identify. The larvae are associated with a variety of food plants including the *Rubiaceae* (coffee family), the *Moraceae* (mulberry family), and the *Urticaceae* (nettles).

ADELPHA BREDOWII
California Sister

Size	Zone	Status
3⅜ in • 85mm	1•2	Not protected

Habitat & Ecology deciduous woodlands; larval food plant: *Fagaceae* (oak family)
Other Characteristics BELOW 3 gray bars edged in forewing cell; gray and orange stripes at base of hindwing; gray spot bands on both wings; FEMALE larger postmedian white band
Subspecies 2 described

ADELPHA LEUCERIA

Size | **Zone** | **Status**
2³/₈ in • 60mm | 2 | Not protected

Habitat & Ecology forest up to 4,500 ft (1,500m); males patrol and have been observed hilltopping
Other Characteristics ABOVE silver gray spot bands on forewing; 3 silver gray bars edged in blackish brown in forewing cell; FEMALE duller
Subspecies 2 described

ADELPHA MELONA

Size | **Zone** | **Status**
2 in • 50mm | 2 | Not protected

Habitat & Ecology open, mesic tropical forest
Other Characteristics BELOW glossy white stripes on both wings; FEMALE less orange on forewing
Subspecies 5 described

Genus Aglais

Found in the northern temperate zones, members of this genus are associated with nettles as larvae. Adults hibernate.

AGLAIS URTICAE
Small Tortoiseshell

Size | **Zone** | **Status**
2 in • 50mm | 3 | Not protected

Habitat & Ecology flowery meadows to 7,000 ft (2,000m); sometimes observed hilltopping
Other Characteristics BELOW brown submarginal spot band on outer hindwing; FEMALE smaller; more yellow on forewing
Subspecies 6 described

Genus Agrias

Comprised of about 10 neotropical species, these robust butterflies are brightly colored and highly prized by collectors.

AGRIAS NARCISSUS

Size	Zone	Status
3³/₈ in • 85mm	2	Not protected

Habitat & Ecology tropical rain forest; attracted to fermenting fruit, carrion, and dung
Other Characteristics BELOW overscaled with gray blue on forewing apex and on hindwing; prominent blue spot band on hindwing with recurved yellow bands; FEMALE larger; blue duller
Subspecies 5 described

Genus Amathuxidia

These tropical butterflies are closely related to *Zeuxidia* and widely distributed in Southeast Asia, Indonesia, and New Guinea.

AMATHUXIDIA AMYTHAON
Koh-I-Noor Butterfly

Size	Zone	Status
4¹/₂ in • 115mm	5•6	Not protected

Habitat & Ecology primary rain forest; grassy plains; attracted to fermenting fruit
Other Characteristics ABOVE violet blue diffuse band extending from costa to tornus; FEMALE above, yellow transverse band on forewing costa
Subspecies 10 described

Genus Amauris

The more than 15 species of this African genus are members of the *Danainae*. These butterflies sequester toxic compounds from *Asclepiadaceae* (hoya) as larvae.

AMAURIS ELLIOTI
Ansorge's Danaid

Size	Zone	Status
3¹/₈ in • 80mm	4	Not protected

Habitat & Ecology open montane woodland and forests; larval food plant: *Asclepiadaceae* (hoya)
Other Characteristics BELOW paler; FEMALE similar
Subspecies 4 described

AMAURIS HECATE
Dusky Danaid, Black Friar

Size	Zone	Status
3¹/₈ in • 80mm	4	Not protected

Habitat & Ecology edges of deep forests
Other Characteristics BELOW more submarginal markings on forewing; outer part of hindwing overscaled with white; veins darkened black; FEMALE larger; forewings rounder
Subspecies 2 described

Genus Amnosia

This variable genus with a single species is restricted to the islands of Indonesia.

AMNOSIA DECORA EUDAMIA

Size	Zone	Status
3 in • 75mm	5	Not protected

Habitat & Ecology highland rain forests; larval food plant: *Elatostema* (waterwort)
Other Characteristics BELOW pale lavender flush and 3 eyespots near forewing apex; 4 prominent eyespots on hindwing; FEMALE similar
Subspecies 7 described

Genus Anaea

With more than 15 species, members of this generally brightly colored, neotropical genus are sexually dimorphic. The pattern below is similar to mottled leaves. Larval food plants include members of the *Euphorbiaceae* (thorn family).

ANAEA ANDRIA
Goatweed Butterfly

Size	Zone	Status
2⁵/₈ in • 65mm	1•2	Not protected

Habitat & Ecology forest edges; by pools; female reclusive and rarely seen
Other Characteristics BELOW white at forewing apex and along wing margins; hindwing darker at base; FEMALE darker; yellow median band on both wings
Subspecies none described

Genus Anartia

This genus is widely distributed in southern Florida, Texas, and Central and South America.

Size	Zone	Status
1⅞ in • 45mm	1•2	Not protected

Habitat & Ecology open scrub; disturbed areas; larval food plant: *Lippia* (oregano)
Other Characteristics BELOW paler; MALE forewing margin more angular; paler; larger
Subspecies 5–6 of doubtful validity described

Genus Antanartia

This genus is restricted to Africa, and is similar to the European *Vanessa*.

ANTANARTIA SCHAENEIA
Long-tailed Admiral

Size	Zone	Status
2⅜ in • 60mm	4	Not protected

Habitat & Ecology associated with montane forests; larval food plant: *Urticaceae* (nettles)
Other Characteristics BELOW white subapical bar on forewing costa; white and lavender on hindwing; FEMALE wings more rounded
Subspecies 3 described

Genus Antirrhea

This neotropical genus composed of 20 species has an unusual lobed posterior forewing and a somewhat angular hindwing. Life history for most species is incomplete, although they appear to be associated with *Palmae* (palms).

ANTIRRHEA AVERNUS

Size	Zone	Status
3⅞ in • 95mm	2	Not protected

Habitat & Ecology dense rain forest
Other Characteristics BELOW cream postmedian band on both wings; hindwing margin edged in yellow; veins darkened; FEMALE similar
Subspecies none described

ANTIRRHEA MILTIADES

Size	Zone	Status
4 in • 100mm	2	Not protected

Habitat & Ecology swamps; dense tropical rain forest; larval food plant: *Geonoma longivaginata* (member of palm family)
Other Characteristics ABOVE 2 or 3 white spots at forewing apex; fine cream spot band on hindwing with posterior 2 spots enlarged; FEMALE lacks prominent lobe on forewing
Subspecies none described

ANTIRRHEA PTEROCOPHA

Size	Zone	Status
4³/₈ in • 110mm	2	Not protected

Habitat & Ecology dark areas of dense tropical rain forest, along forest streams; larval food plant: *Arecaceae* (assai palm)
Other Characteristics BELOW both wings overscaled with pale lavender and white submarginal spots; FEMALE prominent orange tan submarginal patch
Subspecies none described

Genus Apatura

This genus is made up of powerful fliers that are widely distributed. The larval food plants include *Salix* (willow).

APATURA IRIS
Purple Emperor

Size	Zone	Status
3 in • 75mm	3	Not protected

Habitat & Ecology flies in the tops of established woods and along edges of woods
Other Characteristics BELOW white markings enlarged; margins of both wings gray; orange eyespot on forewing; FEMALE; above, less purple
Subspecies several of doubtful validity described

Genus Apaturina

Closely related to *Apatura*, this genus comprises a single species from New Guinea.

APATURINA ERMINIA

Size	**Zone**	**Status**
3⁵/₈ in • 90mm	6	Not protected

Habitat & Ecology mature rain or secondary forest; usually land upside down on tree trunks
Other Characteristics BELOW forewing band white; pupilled eyespot edged in rust above tornus; eyespots encircled in dull gold; FEMALE dimorphic; rust on posterior wings
Subspecies 10 described

Genus Aphantopus

This small genus of woodnymph butterflies comprises possibly 2 species.

Genus Araschnia

Characterized by the net-like patterns below, this genus comprises 1 species, which has 2 color forms: (1) tawny and black above; (2) black with whitish markings above.

APHANTOPUS HYPERANTUS
Ringlet

Size	**Zone**	**Status**
2 in • 50mm	3•5	Not protected

Habitat & Ecology moist grasslands and open woodlands up to 5,000 ft (1,500m)
Other Characteristics ABOVE brown; 2 dark eyespots on both wings with those of hindwing larger; cream fringe on wing margins; FEMALE similar
Subspecies several of doubtful validity described

ARASCHNIA LEVANA
Map Butterfly

Size	**Zone**	**Status**
1⁵/₈ in • 40mm	3	Not protected

Habitat & Ecology open woodlands; larval food plant: *Urticaceae* (nettles)
Other Characteristics ABOVE margins pale yellow with veins darkened; FEMALE similar
Subspecies none described

Genus Argynnis

These are the Silverspots or Large Fritillaries and are widely distributed in Europe and Asia.

ARGYNNIS ADIPPE
High Brown Fritillary

Size	Zone	Status
2³/₈ in • 60mm	3	Not protected

Habitat & Ecology heaths; woods; high montane meadows; larval food plant: *Viola* (violet)
Other Characteristics BELOW forewing apex and margins overscaled with dull olive green; olive green along hindwing margin; FEMALE above, duller in coloration; larger
Subspecies fewer than 30 described so far

ARGYNNIS PAPHIA
Silver-washed Fritillary

Size	Zone	Status
2⁵/₈ in • 65mm	3	Not protected

Habitat & Ecology woodland
Other Characteristics BELOW overscaled on both wings with chartreuse green; lavender white postmedian and marginal bands, and 2 faint postdiscal spot bands on hindwing; FEMALE larger; duller; above, lacks androconial patches
Subspecies 2 described

Genus Argyrophenga

Restricted to New Zealand, this satyrid genus has some unusual silver stripes below.

ARGYROPHENGA ANTIPODUM

Size	Zone	Status
2¹/₈ in • 55mm	6	Not protected

Habitat & Ecology subalpine meadows
Other Characteristics ABOVE large orange postmedian patch; black eyespots with bluish pupils present on both wings; FEMALE similar
Subspecies none described

Nymphalidae

Genus Argyrophorus

As the common name implies, these neotropical butterflies have silver on the upper wing surface.

ARGYROPHORUS ARGENTEUS
Silver Butterfly

Size	Zone	Status
2^1/$_8$ in • 55mm	2	Not protected

Habitat & Ecology Andean high desert up to 8,500 ft (2,600m); fast fliers; avid flower visitors
Other Characteristics BELOW black apical, pupilled eyespot on forewing; hindwing overscaled with blackish gray; postdiscal and marginal bands evident; FEMALE darker margins, especially on forewing; dark spots in the forewing cell
Subspecies none described

Genus Asterocampa

The 5 species of this genus are commonly called the Hackberry butterflies because of their larval foodplant *Celtis* (hackberry).

ASTEROCAMPA CELTIS
Hackberry Butterfly, Hackberry Emperor

Size	Zone	Status
2^3/$_8$ in • 60mm	1•2	Not protected

Habitat & Ecology open woodland; forest edges
Other Characteristics BELOW pupilled submarginal band on hindwing; FEMALE similar
Subspecies 8–10 described

Genus Aterica

This is a small genus of African and Madagascan species.

ATERICA GALENE
Forest Glade Nymph

Size	Zone	Status
3^1/$_8$ in • 80mm	4	Not protected

Habitat & Ecology clearings; rain forest margins
Other Characteristics BELOW forewing apex and entire hindwing overscaled with cream-buff; lighter in hindwing cell; FEMALE hindwing patch white or orange
Subspecies 3 described

Genus Baeotus

A neotropical genus of 2 or 3 species, this group is characterized by the unique wing margins similar to *Charaxes*.

BAEOTUS BAEOTUS

Size
3⁵/₈ in • 90mm

Zone
2

Status
Not protected

Habitat & Ecology infrequently encountered in tropical rain forest; larval food plant: unknown
Other Characteristics BELOW cream ground color; intricate pattern of short bluish black bars; orange blotches on forewing; FEMALE blue markings replaced by dull yellow
Subspecies 1 described

Genus Basilarchia

Closely related to the Eurasian genus *Limenitis*, these butterflies are commonly called the Admirals and associated with a wide variety of tree species as larvae. This group mimics various danaids (Monarch, Queen).

BASILARCHIA LORQUINI
Lorquin's Admiral, Orange Tip Admiral

Size
2⁷/₈ in • 70mm

Zone
1

Status
Not protected

Habitat & Ecology open highland woods; larval food plants: include *Salix* (willow)
Other Characteristics BELOW gray violet bars at hindwing base; gray violet submarginal spot bands on both wings; FEMALE similar
Subspecies 3 described

BASILARCHIA ARCHIPPUS
Viceroy

Size
3 in • 75mm

Zone
1•2

Status
Not protected

Habitat & Ecology open woodlands; edges of woods; larval food plant: *Salix* (willow)
Other Characteristics BELOW paler; especially at forewing subapex and on hindwing; FEMALE similar
Subspecies 6 described

Genus Bassarona

This genus of rather large butterflies is similar to *Euthalia* and is restricted to Southeast Asia and the Molucca Islands.

BASSARONA DUDA

Size	Zone	Status
4¹/₈ in • 105mm	5	Not protected

Habitat & Ecology rain forest; forest edges; little known about life history
Other Characteristics BELOW suffused with dark olive brown along outer margins of both wings; FEMALE similar
Subspecies 2 described

Genus Bicyclus

This is an African genus of brown butterflies composed of more than 132 species.

BICYCLUS SAFITZA
Common Bush Brown

Size	Zone	Status
2 in • 50mm	4	Not protected

Habitat & Ecology common in rain forests, open glades and along paths; settles on leaves on the ground; larval food plant: *Poaceae* (grass family)
Other Characteristics BELOW 2 faint eyespots near apex of forewing; FEMALE duller
Subspecies none described

Genus Biblis

There is only 1 species in this genus that is widely distributed from south Texas to Argentina.

BIBLIS HYPERIA
Red Rim

Size	Zone	Status
2⁵/₈ in • 65mm	1•2	Not protected

Habitat & Ecology clearings in dense to open woodlands and mesic tropical forests; larval food plant: *Tragia volubilis* (nose burn)
Other Characteristics BELOW duller; hindwing spot band smaller and pinker; FEMALE similar
Subspecies several of doubtful validity described

Genus Boloria

This is a genus of subalpine and subarctic fritillaries.

BOLORIA AQUILONARIS
Cranberry Fritillary

Size
1³/₈ in • 35mm

Zone
3

Status
Not protected

Habitat & Ecology bogs and moors; larval food plant: *Vaccinium uliginosum* (bog bilberry)
Other Characteristics BELOW 7 silver marginal spots on hindwing; FEMALE duller; darker
Subspecies none described

BOLORIA NAPAEA
Mountain Fritillary, Napea Fritillary

Size
1⁵/₈ in • 40mm

Zone
1•3

Status
Not protected

Habitat & Ecology lush mountain slopes; montane meadows; larval food plants: *Polygonum* (bistorts)
Other Characteristics BELOW forewing apex overscaled with white and lavender; FEMALE darker areas more extensive; olive green at base
Subspecies 10 described

Genus Brenthis

These are smaller fritillaries from Europe and Asia with more rounded wings than *Boloria*.

BRENTHIS INO
Lesser Marbled Fritillary

Size
1⁵/₈ in • 40mm

Zone
3

Status
Not protected

Habitat & Ecology boreal forests; swampy meadows and bogs; larval food plants: include *Rubus idaea* (raspberry), *Filpendula ulmaria* (meadowsweet)
Other Characteristics BELOW forewing apex rust bar edged with yellow; hindwing veins rust, with yellow median band, and overscaled with golden yellow along margin; FEMALE similar
Subspecies 10 described

Genus Brintesia

This is a genus of large browns that are found in open woodlands.

▲

BRINTESIA CIRCE
Great Banded Grayling

Size	Zone	Status
2⁵/₈ in • 65mm	3	Not protected

Habitat & Ecology open woodlands and clearings up to about 4,500 ft (1,500m); larval food plant: *Poaceae* (grass family)
Other Characteristics ABOVE has single apical ocellus; FEMALE larger; blackish; more suffused white median bands across wings above
Subspecies none described

Genus Byblia

These 2 species of African butterfly are associated with woodlands and mesic scrub habitats.

BYBLIA ILITHYIA
Joker

Size	Zone	Status
2¹/₈ in • 55mm	4	Not protected

Habitat & Ecology savannah; scrubby woodland; larval food plant: *Euphorbiaceae* (nettles)
Other Characteristics BELOW fine white chevrons on forewing; FEMALE larger
Subspecies none described

Genus Caerois

Closely related to the genus *Morpho*, there are 2 species in this genus.

CAEROIS CHORINAEUS

Size	Zone	Status
4 in • 100mm	2	Not protected

Habitat & Ecology deep in rain forest and woods
Other Characteristics BELOW prominent androconial patch on forewing; fine white submarginal spots on hindwing; FEMALE above, lacks androconial markings
Subspecies none described

Genus Caligo

Commonly called owl butterflies because of their underside wing pattern and markings, these neotropical butterflies fly at dawn and at dusk along rain forest paths.

CALIGO BELTRAO
Ornate Owl Butterfly

Size	**Zone**	**Status**
5⅝ in • 140mm	2	Not protected

Habitat & Ecology woods; forest clearings
Other Characteristics BELOW mottled "owl's head" appearance; FEMALE similar
Subspecies none described

CALIGO MEMNON
Owl Butterfly

Size	**Zone**	**Status**
6 in • 150mm	2	Not protected

Habitat & Ecology open clearings in forests and banana groves where they may be a pest; other larval food plant: *Heliconia* (wild plantain)
Other Characteristics ABOVE similar; FEMALE gray brown above with darker hindwings
Subspecies 3 described

Genus Cercyonis

The 4 species of this North American group have enlarged eyespots on the forewings, and other distinct structures.

CERCYONIS MEADII
Mead's Wood Nymph

Size	**Zone**	**Status**
2¼ in • 60mm	1	Not protected

Habitat & Ecology mesic meadows; open woodlands; canyons and sagebrush plains; larval food plant: *Poaceae* (grasses)
Other Characteristics ABOVE hindwing uniform brown; FEMALE similar
Subspecies 4 described

Nymphalidae

CERCYONIS PEGALA
Blue-eyed Grayling, Goggle Eye, Common Wood Nymph

Size	**Zone**	**Status**
3 in • 75mm	1	Not protected

Habitat & Ecology open woodlands; meadows; grasslands; larval food plant: *Poaceae* (grasses)
Other Characteristics ABOVE brown; forewing ocelli either in a yellow or brown field; FEMALE ocelli enlarged
Subspecies more than 15 described

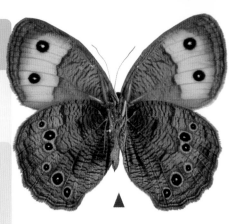

Genus Cethosia
These are forest-dwelling species that feed on *Passifloraceae* (violales) as larvae. All 10–20 recognized species are tropical, and restricted to Asian or Australian regions.

♀

CETHOSIA NIETNERI

Size	**Zone**	**Status**
4¹/₈ in • 105mm	5	Not protected

Habitat & Ecology open forest clearings
Other Characteristics BELOW reduced black markings; golden orange submarginal bands on both wings; MALE similar
Subspecies 2 described

CETHOSIA CYANE

Size	**Zone**	**Status**
4 in • 100mm	5	Not protected

Habitat & Ecology up- and lowland rain forests
Other Characteristics BELOW 3 blue bars outlined in black on forewing cell; FEMALE dusky white
Subspecies 2 from Burma and India described

Genus Charaxes

The more than 100 species in this large genus are primarily distributed in Africa, with 1 species in Europe and 12 in Asia. *Charaxes* are noted for their remarkable coloration, especially the intense patterns and colors below, and thus are favorites of collectors. Most species have 2 pairs of tails, are very strong, powerful fliers, and are often engaged in hilltopping. They are attracted to fermenting fruit, sap flows, and carrion.

CHARAXES CASTOR
Giant Charaxes

Size	Zone	Status
4³⁄₈ in • 110mm	4	Not protected

Habitat & Ecology woodlands and associated brush; males terrltorial and hilltop
Other Characteristics ABOVE blackish brown ground color; FEMALE larger forewing bands
Subspecies none described

CHARAXES BOHEMANI
Large Blue Charaxes

Size	Zone	Status
4¹⁄₈ in • 105mm	4	Not protected

Habitat & Ecology dry deciduous forest; fringes of evergreen forests; males very territorial; larval food plant: *Fabaceae* (pea family)
Other Characteristics BELOW olive gray ground color; fine white and black lines on hindwing; FEMALE duller; with a broad white band between the black and blue areas of the forewing
Subspecies none described

CHARAXES CITHAERON
Blue-spotted Charaxes

Size	Zone	Status
3⁵⁄₈ in • 90mm	4	Not protected

Habitat & Ecology clearings and paths in open to high forests and savannahs
Other Characteristics BELOW olive gray with faint orange submarginal spot; FEMALE larger; above, brown basally; white recurved band at apex and 2 black spots; hindwing lavender and black with a lavender submarginal spot band
Subspecies 5 described

CHARAXES DRUCEANUS
Silver-barred Charaxes

Size	**Zone**	**Status**
3³/₈ in • 85mm	4	Not protected

Habitat & Ecology forest paths and clearings
Other Characteristics BELOW linear markings outlined in white; blue bands on hindwing; FEMALE larger; duller; with rounded wings; more bluish white toward hindwing tornus
Subspecies 12 described

CHARAXES EURIALUS

Size	**Zone**	**Status**
4 in • 100mm	6	Not protected

Habitat & Ecology open woodlands and montane forests; larval food plants: *Myrtaceae* (myrtle family), *Melianthaceae* (honeybush)
Other Characteristics BELOW dark red brown; prominent blue eyespot on forewing; FEMALE distinct orange forewing; duller
Subspecies none described

CHARAXES HANSALI
Cream-banded Charaxes

Size	**Zone**	**Status**
3¹/₄ in • 85mm	4	Not protected

Habitat & Ecology montane scrub; open woodland; larval food plants: *Salvadoraceae* (salvadora family)
Other Characteristics BELOW dull olive rust; glossy white bands; basal black markings outlined in blue; FEMALE longer hindwing tails
Subspecies 4 described

▲
♀

CHARAXES JASIUS
Foxy Charaxes, Two-tailed Pasha

Size	**Zone**	**Status**
3³/₈ in • 85mm	3•4	Not protected

Habitat & Ecology forest clearings; forest edges
Other Characteristics ABOVE forewing margin suffused with orange; MALE orange submarginal spotband on hindwing
Subspecies 6 described, all but one from Africa

CHARAXES NITEBIS

Size	**Zone**	**Status**
4 in • 100mm	6	Not protected

Habitat & Ecology open montane forest
Other Characteristics BELOW light brown ground color; blue capped submarginal spot band on hindwing; FEMALE similar
Subspecies 3 described

Genus Chazara

This is a genus of browns that occur in the Himalayan foothills. They are characterized by distinct wing maculation patterns.

CHAZARA HEYDENREICHII

Size	**Zone**	**Status**
2¹/₈ in • 55mm	5	Not protected

Habitat & Ecology open rocky montane areas
Other Characteristics BELOW similar; FEMALE with a more golden yellow ground color
Subspecies none described

Genus Chlosyne

This is a fairly diverse genus comprising more than 20 species. It is derived originally from *Melitaea* and includes the patch butterflies.

♀

CHLOSYNE PALLA
Creamy Checkerspot

Size	Zone	Status
1⁷/₈ in • 45mm	1	Not protected

Habitat & Ecology open montane woods; forests; scrub areas; larval food plant: *Castilleja breviflora* (member of the figwort family)
Other Characteristics ABOVE dark blackish brown markings; MALE smaller; can be darker
Subspecies 6 described

♀

CHLOSYNE THEONA
Mexican Checkerspot

Size	Zone	Status
1⁷/₈ in • 45mm	1•2	Not protected

Habitat & Ecology dry canyon washes and sparse woodlands; larval food plant: *Castilleja* (figwort family) and *Verbenaceae* (verbena)
Other Characteristics BELOW paler; cream spot bands on hindwing; MALE color patterns brighter
Subspecies 3 described

CHLOSYNE CALIFORNICA
California Patch

Size	Zone	Status
1³/₄ in • 40mm	1	Rare

Habitat & Ecology deserts and canyon washes; larval food plant: *Viguiera deltoides* (golden eye)
Other Characteristics BELOW equal sized white spots on hindwing; FEMALE similar
Subspecies none described

Genus Cirrochroa
This genus of Asian butterflies is composed of 20 species that are characterized by rather large wings with faintly scalloped margins. The antennal club is black. These butterflies have slow flight, occasionally gliding.

CIRROCHROA REGINA

Size
3³/₈ in • 85mm

Zone
6

Status
Not protected

Habitat & Ecology primary rain forest; secondary forest; larval food plants: *Flacourtiaceae* (Flacourtiaceae), *Euphorbiaceae* (thorn family)

Other Characteristics BELOW duller; FEMALE below, patterned and suffused with violet
Subspecies 7 described

CIRROCHROA THULE

Size
4 in • 100mm

Zone
6

Status
Not protected

Habitat & Ecology mature secondary and primary rain forest

Other Characteristics BELOW base of wings overscaled with iridescent gray; FEMALE duller
Subspecies 2 described

Genus Cithaerias
A neotropical genus of 10 species of diaphanous butterfly, these are characterized by discrete patterns. They fly in the understory of tropical rain forests, but little information is available on larval food plants or life history.

CITHAERIAS PHILIS

Size
2⁵/₈ in • 65mm

Zone
2

Status
Not protected

Habitat & Ecology understory of primary tropical rain forest
Other Characteristics BELOW similar; FEMALE less lavender shading in hindwing
Subspecies none described

Nymphalidae

CITHAERIAS SONGOANA

Size	Zone	Status
3 in • 75mm	2	Not protected

Habitat & Ecology understory of primary tropical rain forest
Other Characteristics BELOW similar; FEMALE the reddish fuchsia patch of hindwing is edged in brown, including the eyespot
Subspecies none described

Genus Clossiana

This genus of small fritillaries comprises 20 species and is closely related to *Boloria*. Many are endemic to areas north of the Arctic Circle, and all are distributed in the northern temperate zone.

CLOSSIANA BELLONA
Meadow Fritillary

Size	Zone	Status
2 in • 50mm	1	Not protected

Habitat & Ecology moist meadows, fields, and along streams adjacent to disturbed habitats; larval food plant: *Viola* (violet)
Other Characteristics ABOVE similar; heavier postmedian bands; FEMALE slightly larger
Subspecies 3 described

CLOSSIANA CHARICLEA
Arctic Fritillary

Size	Zone	Status
1³/₈ in • 35mm	1•3	Not protected

Habitat & Ecology arctic tundra; one of the few butterflies that occurs in Greenland; larval food plants: *Salix* (willow), *Polygonum* (bistorts)
Other Characteristics ABOVE heavier markings; MALE bright orange above with dark markings
Subspecies 3 described

CLOSSIANA EUPHRYOSYNE
Pearl-bordered Fritillary

Size	Zone	Status
1⁷/₈ in • 45mm	3	Not protected

Habitat & Ecology open alpine woods up to 6,000 ft (2,000m); larval food plant: *Viola* (violet)
Other Characteristics ABOVE heavier black brown markings; FEMALE more dusky
Subspecies 4 described

CLOSSIANA FRIGGA
Willow-bog Fritillary, Frigga's Fritillary

Size	Zone	Status
1⁷/₈ in • 45mm	1•3	Not protected

Habitat & Ecology willow and black spruce bogs; larval foodplants: *Rubus chamaemorous* (cloudberry), *Salix* (willow)
Other Characteristics BELOW hindwing with broad violet band on outer margin and prominent white patch; FEMALE similar; duller
Subspecies 4 described

Genus Coelites

This is a satyrid genus comprising 2 species that are associated with the dense lowland forests of Asia and Australia.

COELITES EPIMINITHIA

Size	Zone	Status
3¹/₈ in • 80mm	5	Not protected

Habitat & Ecology lowland primary rain forest
Other Characteristics ABOVE purple patches on both wings; black androconial patch on inner margin of hindwing; FEMALE purple reduced
Subspecies 3 described

Genus Coenonympha

This is a large genus of browns or heaths that occur in the northern temperate zone and are associated with various grasses as larvae.

COENONYMPHA HAYDENII
Wyoming Ringlet, Yellowstone Ringlet

Size	Zone	Status
2 in • 50mm	1	Protected

Habitat & Ecology grassy meadows at moderate elevations; life history is unknown
Other Characteristics ABOVE pale uniform brown; gray brown fringes on hindwing; FEMALE similar
Subspecies none described

COENONYMPHA CORINNA
Corsican Heath

Size	Zone	Status
1¹/₈ in • 30mm	3	Not protected

Habitat & Ecology open grassy meadows up to 3,000 ft (1,000m)
Other Characteristics BELOW paler; ocelli encircled in yellow; FEMALE above, color brighter
Subspecies 2 described

COENONYMPHA PAMPHILUS
Small Heath

Size	Zone	Status
1¹/₈ in • 30mm	3	Not protected

Habitat & Ecology roadsides, grassy fields, and meadows up to 6,000 ft (2,000m)
Other Characteristics ABOVE gray brown fringes on hindwing; FEMALE above, duller and darker
Subspecies 6 described

Genus Corades

The 15 neotropical, montane species of this genus are associated with bamboo forests. The tails and other structural features make them unmistakable.

CORADES IDUNA

| **Size** | **Zone** | **Status** |
| 3 in • 75mm | 2 | Not protected |

Habitat & Ecology montane cloud forests
Other Characteristics BELOW brown to mottled brown overscaled with white; FEMALE orange patch on hindwing paler and more diffuse
Subspecies 3 described

Genus Cymothoe

This large African genus, with more than 72 species, exhibits marked sexual dimorphism. These butterflies are rapid fliers, quite territorial and pugnacious, and are favorites of collectors.

CYMOTHOE FUMANA
Gold-banded Glider

| **Size** | **Zone** | **Status** |
| 3 in • 75mm | 4 | Not protected |

Habitat & Ecology rain forest and forest edges
Other Characteristics BELOW mottled brown with purplish overcast; FEMALE chocolate brown on the basal two-thirds of wings with distal third covered in orange brown
Subspecies 2 described

CYMOTHOE SANGARIS
Red Glider, Blood Red Cymothoe

| **Size** | **Zone** | **Status** |
| 2⁷/₈ in • 70mm | 4 | Not protected |

Habitat & Ecology rain forest canopy
Other Characteristics BELOW reddish brown ground color; darker postmedian bands; FEMALE above, tan white; broken beige bands and markings on the outer half of wings
Subspecies 2 described

Genus Cynandra

This genus of African butterflies comprises a single species that flies in the deep forests. Sexes are dimorphic.

CYNANDRA OPIS
Shining Blue Nymph, Blue Banded Nymph

Size 2 1/8 in • 55mm
Zone 4
Status Not protected

Habitat & Ecology rain forest shade
Other Characteristics BELOW dark brown; diffused white markings; FEMALE brown ground color; above, yellow markings
Subspecies none described

Genus Cynthia

These butterflies are highly migratory and adapted as larvae to a wide variety of food plants.

Genus Cyrestis

Members of this genus are known as map butterflies. They are widely distributed in Asia and Australia with one species in Africa. Species are often associated with *Ficus* (fig) as larvae.

CYNTHIA CARDUI
Painted Lady, Cosmopolite

Size 2 7/8 in • 70mm
Zone 1•2•3•4•5
Status Not protected

Habitat & Ecology variable because of its strong migratory tendencies, but not deep rain forests; larval foodplants: *Cirsium* (bullthistle)
Other Characteristics BELOW forewing pink with gray brown at apex; FEMALE paler; wings rounder
Subspecies none described

CYRESTIS ACHAETES
Map Butterfly

Size 2 1/8 in • 55mm
Zone 6
Status Not protected

Habitat & Ecology open clearings, along paths in lowland rain forests; can be gregarious; larval food plant: *Malaisia scandens* (burney vine)
Other Characteristics BELOW paler; FEMALE similar
Subspecies 3 described

CYRESTIS CAMILLUS
African Map Butterfly

Size	Zone	Status
2¹/₈ in • 55mm	4	Not protected

Habitat & Ecology rain forest edges especially near Kakamega, Kenya
Other Characteristics BELOW markings reduced; basal markings duller; FEMALE similar
Subspecies 3 described

CYRESTIS MAENALIS

Size	Zone	Status
2³/₈ in • 60mm	5	Not protected

Habitat & Ecology rain forest, near streams
Other Characteristics BELOW paler; FEMALE similar
Subspecies 11 described

Genus Danaus

Butterflies of this genus, that includes the Monarch, are generally known as the tiger butterflies. They are associated with members of the *Asclepiadaceae* (hoya) as larvae and are toxic to predators.

DANAUS AFFINIS

Size	Zone	Status
3 in • 75mm	6	Not protected

Habitat & Ecology open clearings; along edges of forests; lowland coastal areas; beach strands
Other Characteristics BELOW gray white overscaling on hindwing; FEMALE lacks scent patch on hindwing
Subspecies 8 described

DANAUS CHRYSIPPUS AEGYPTIUS
Plain Tiger, African Monarch,
Lesser Wanderer, Golden Danaid

Size
3¹/₈ in • 80mm

Zone
4•5•6

Status
Not protected

Habitat & Ecology open woodlands; fields; highly migratory
Other Characteristics BELOW more dull orange on hindwing; MALE forewing apices more acute; above, hindwing androconial scent patch
Subspecies more than 24 described

DANAUS PARANTICA ASPASIA

Size
3¹/₈ in • 80mm

Zone
5•6

Status
Not protected

Habitat & Ecology various, from open woodlands to forests at low and moderate elevations; larval food plant: *Gymnema sylvestre* (Gymnema)
Other Characteristics BELOW larger submarginal spot bands; FEMALE duller and paler
Subspecies 12 described

DANAUS GILIPPUS
Queen

Size
3⁵/₈ in • 90mm

Zone
1•2

Status
Not protected

Habitat & Ecology open areas; edges of woods; migratory yet territorial; larval food plants: *Asclepiadaceae* (hoya), including *Sarcostemma viminale* (sacred soma)
Other Characteristics BELOW larger white submarginal spot bands; FEMALE lacks prominent androconial patch on hindwing
Subspecies 8 described

DANAUS PLEXIPPUS
Milkweed, Monarch

Size	**Zone**	**Status**
4 in • 100mm	1-6	Not protected*

Habitat & Ecology open areas; clearings; forest edges; gardens
Other Characteristics BELOW similar; FEMALE more brown; lacks prominent scent patch
Subspecies 5 described

Genus Diagora

This genus is composed of a single species and is similar to a number of Asian danaids.

Genus Dione

These neotropical butterflies are noted for their contrasting coloration and hindwing silver spots. They are closely related to the genus *Heliconius*.

DIAGORA MENA

Size	**Zone**	**Status**
3⁷/₈ in • 95mm	5	Not protected

Habitat & Ecology open clearings in widely spaced secondary forests and along edges of woods
Other Characteristics BELOW similar; FEMALE similar
Subspecies none described

DIONE MONETA POEYI
Mexican Silverspot,
Mexican Silver-spotted Fritillary

Size	**Zone**	**Status**
3 in • 75mm	1•2	Not protected

Habitat & Ecology open country; woodland; dry tropical forest edges; solitary and migratory; larval food plant: *Passiflora* (passion flower)
Other Characteristics ABOVE similar; FEMALE above, reddish brown with black markings; veins outlined in black on both wings
Subspecies 2 described

Genus Doleschallia

Totalling 6 species, members of this genus are strong fliers and forest dwellers.

DOLESCHALLIA DASCYLUS

Size	Zone	Status
$3^5/_8$ in • 90mm	6	Not protected

Habitat & Ecology along paths and edges of dense woods or in marginal rain forest; larval food plants: *Acanthaceae* (acanthia family)
Other Characteristics BELOW 2 prominent eyespots on hindwing; FEMALE forewing bands slightly larger
Subspecies 4 described

Genus Doxocopa

The 30 neotropical butterflies in this genus are strong fliers and markedly sexually dimorphic.

Genus Drucina

The 3 brown species in this genus are associated with cloud forests.

DOXOCOPA LAURENTIA CHERUBINA

Size	Zone	Status
$2^5/_8$ in • 65mm	2	Not protected

Habitat & Ecology edges and slopes of tropical rain forest; larval food plant: *Celtis* (hackberry)
Other Characteristics BELOW tan at base fading to silvery gray at apex; FEMALE above, brown with an orange patch at forewing apex and a white band across forewing
Subspecies 2 described

DRUCINA CHAMPIONI

Size	Zone	Status
4 in • 100mm	2	Not protected

Habitat & Ecology cloud forests; forest edges; larval food plant: *Poaceae* (bamboo)
Other Characteristics BELOW white at forewing apex; hindwing veins darkened; FEMALE similar
Subspecies none described

DRUCINA LEONATA

Size	**Zone**	**Status**
3⅝ in • 90mm	2	Not protected

Habitat & Ecology deep recesses of montane cloud forest; adults rest at base of trees; larval food plant: *Poaceae* (bamboo)
Other Characteristics BELOW mottled appearance; veins darkened; fine white submarginal spots FEMALE similar
Subspecies none described

Genus Elymnias

These butterflies are known as the Palmflies because of their food plants as larvae. They have unusual wing margins and mimic butterflies from other families.

Genus Dryas

Closely related to *Heliconius*, this genus comprises a single species that is variable throughout its range.

DRYAS IULIA
Julia

Size	**Zone**	**Status**
3⅝ in • 90mm	1•2	Not protected

Habitat & Ecology fast flier in open woodlands and forests, along paths and forest margins; flower visitor; larval food plant: *Passiflora* (passion flower)
Other Characteristics BELOW purplish tan to light brown; FEMALE duller; median tan bands on both wings
Subspecies more than 15 described

ELYMNIAS AGONDAS
Palmfly

Size	**Zone**	**Status**
3⅝ in • 90mm	5•6	Not protected

Habitat & Ecology rain forest clearings, marginal secondary forests; larval food plants: *Arecaceae* (assai palm), *Musaceae* (banana)
Other Characteristics ABOVE more vivid; MALE dimorphic; below, black with similar marks; above, black with 3 eyespots on hindwing, and white hindwing patch
Subspecies 12 described

ELYMNIAS CASIPHONE MALELAS

Size	Zone	Status
4 in • 100mm	5	Not protected

Habitat & Ecology rain forest paths and edges; larval food plant: *Arecaceae* (assai palm)
Other Characteristics BELOW similar; FEMALE similar
Subspecies 8 described

ELYMNIAS DARA BENGENA

Size	Zone	Status
2⁷/₈ in • 70mm	5	Not protected

Habitat & Ecology rain forest; larval food plant: *Palmae* (palm family)
Other Characteristics BELOW tan scrawls on forewing disk; paler at apex and along margin; FEMALE duller coloration
Subspecies 6 described

ELYMNIAS HYPERMNESTRA
Common Palmfly

Size	Zone	Status
3¹/₈ in • 80mm	5	Not protected

Habitat & Ecology variable, from rain forest to open areas; shade dwellers; larval food plant: *Arecaceae* (assai palm)
Other Characteristics BELOW dark brown scrawls and lines; white spot on hindwing; MALE blue band on forewing may be present or absent; rust band of hindwing more extensive
Subspecies more than 4 described

ELYMNIAS MIMALON

Size	Zone	Status
4 in • 100mm	6	Not protected

Habitat & Ecology open areas and tropical forests; larval food plant: *Arecaceae* (assai palm)
Other Characteristics BELOW tan with darker scrawls; small white mid-costal spot on hindwing FEMALE above, dull dark brown with lilac bands across both wings
Subspecies 2 described

ELYMNIAS NELSONI

Size	Zone	Status
3¹/₈ in • 80mm	5	Rare

Habitat & Ecology rare in deep tropical rain forest of the Malay peninsula
Other Characteristics BELOW dark brown; paler scrawls on both wings; FEMALE duller
Subspecies none described

Genus Elzunia

This small genus of subandean longwings use *Solanaceae* (nightshade family) as larvae, and are consequently distasteful to predators.

ELZUNIA BONPLANDII

Size	Zone	Status
4 in • 100mm	2	Not protected

Habitat & Ecology tropical rain forest
Other Characteristics ABOVE lacks orange patches; yellow submarginal band on both wings; FEMALE similar
Subspecies 2 described

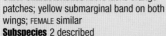

Genus Erebia

This genus of browns comprises more than 100 species that are widespread throughout the higher elevations and in the Arctic. They are generally recognized by their dark coloration— a combination of dark brown, gray, and russet.

EREBIA CALLIAS
Colorado Alpine, Relict Gray Alpine

Size	Zone	Status
2 in • 50mm	1	Protected

Habitat & Ecology moist alpine meadows
Other Characteristics ABOVE dark brown with forewing eyespots; FEMALE similar
Subspecies none described

♀

EREBIA AETHIOPS
Scotch Argus

Size	Zone	Status
2 in • 50mm	3•5	Not protected

Habitat & Ecology subalpine woods
Other Characteristics ABOVE ocelli on hindwing encased in reddish orange; FEMALE paler
Subspecies none described

EREBIA DISA STECKERI
White-spot Alpine, Arctic Ringlet

Size	Zone	Status
2 in • 50mm	1•3	Not protected

Habitat & Ecology arctic or subarctic grasslands or open woods
Other Characteristics ABOVE brown more uniform; lacks ocelli; checkered gray brown fringes; MALE below, darker; less contrasting markings
Subspecies 6 described

EREBIA FASCIATA
Banded Alpine, White-banded Alpine

Size	Zone	Status
2¹/₈ in • 55mm	1•3	Not protected

Habitat & Ecology tundra in Canadian and Alaskan arctic
Other Characteristics ABOVE dark brown with diffuse reddish postmedian forewing patches; FEMALE paler
Subspecies 5 described

EREBIA HERSE

Size	Zone	Status
2¹/₈ in • 55mm	3	Not protected

Habitat & Ecology high alpine meadows
Other Characteristics BELOW similar; FEMALE paler
Subspecies none described

EREBIA MAGDALENA
Magdalena Alpine, Rockside Alpine

Size	Zone	Status
2⁵/₈ in • 65mm	1•3	P (National Parks only)

Habitat & Ecology locally common on montane rock slides well above timberline; larval food plant: *Poaceae* (grasses)
Other Characteristics ABOVE variable, has complete hindwing submarginal spot band; FEMALE similar
Subspecies 5 described

Nymphalidae

▲
♀

EREBIA POLARIS
Arctic Woodland Ringlet

Size	Zone	Status
1³/₄ in • 40mm	3	Not protected

Habitat & Ecology subarctic open grasslands
Other Characteristics ABOVE similar; MALE above markings much brighter and darker
Subspecies none described

▲

EREBIA ROSSII
Arctic Alpine, Ross's Alpine,
Two-dot Alpine

Size	Zone	Status
1⁷/₈ in • 45mm	1	Not protected

Habitat & Ecology tundra; barren country
Other Characteristics ABOVE uniform brown; 2 forewing ocelli encircled in reddish orange; FEMALE similar
Subspecies 4 described

EREBIA THEANO
Theano Alpine, Banded Alpine

Size	Zone	Status
1³/₄ in • 40mm	1•3	Not protected

Habitat & Ecology bogs; moist areas
Other Characteristics BELOW similar; FEMALE paler
Subspecies 8 described

Genus Erebiola

This New Zealand genus is composed of a single species that resembles, but is not closely related to, *Erebia*.

EREBIOLA BUTLERI

Size	Zone	Status
2¹/₈ in • 55mm	6	Not protected

Habitat & Ecology montane meadows
Other Characteristics ABOVE uniformly blackish brown; more prominent forewing eyespots; FEMALE similar
Subspecies none described

Genus Eryphanis

This genus comprises 6 neotropical species and is closely related to *Caligo*, the Owl butterflies. They are associated with *Poaceae* (bamboo) as larvae.

ERYPHANIS AESACUS

Size	Zone	Status
5 in • 120mm	2	Not protected

Habitat & Ecology clearings in cloud forests; dense rain forest
Other Characteristics ABOVE dark brown; dark purple postmedian bands; darker apical spot band on forewing; FEMALE muted; paler
Subspecies 2 described

Genus Eteona

This neotropical genus has a single species with unusual wing patterns.

ETEONA TISIPHONE

Size	Zone	Status
2¹/₈ in • 55mm	2	Not protected

Habitat & Ecology montane Brazilian rain forests
Other Characteristics ABOVE black with broad yellow patch on hindwing; FEMALE dimorphic and polymorphic; similar forewing patterns, or brown with orange forewing spots and hindwing bases
Subspecies several described but are forms of the same species

Genus Ethope

This is a genus of large Asian brown butterflies.

ETHOPE DIADEMOIDES

Size
3³/₈ in • 85mm

Zone
5

Status
Not protected

Habitat & Ecology rain forest
Other Characteristics BELOW paler; prominent eyespot near hindwing apex; FEMALE duller
Subspecies none described

ETHOPE HIMACHALA

Size
4 in • 100mm

Zone
5

Status
Not protected

Habitat & Ecology rain forest
Other Characteristics BELOW paler; duller; MALE darker; lacks white along forewing costa
Subspecies none described

Genus Eunica

This large genus has more than 60 species, primarily from the American tropics. They are generally forest dwellers.

EUNICA NORICA

Size
2³/₈ in • 60mm

Zone
2

Status
Not protected

Habitat & Ecology cloud forests from 200 to 4,000 ft (60–1,200m)
Other Characteristics BELOW mottled brown; double pupilled ocellus near end cell; FEMALE brown with white markings on forewing
Subspecies none described

Genus Euphydryas

These medium-sized butterflies are commonly called the checkerspots of North America and the smaller fritillaries of Europe.

EUPHYDRYAS CHALCEDONA
Chalcedon or Western Checkerspot

Size	**Zone**	**Status**
2 in • 50mm	1	Not protected

Habitat & Ecology open areas; sagebrush flats
Other Characteristics BELOW forewing suffused with reddish orange at base; hindwing dark brown with reddish spots; FEMALE larger; wings rounder
Subspecies more than 17 described

EUPHYDRYAS AURINIA
Marsh Fritillary

Size	**Zone**	**Status**
1³/₄ in • 40mm	3	Not protected

Habitat & Ecology varied, grassy banks; flowery meadows; marsh and bog dweller; larval food plant: *Plantago major* (plantain)
Other Characteristics BELOW yellow gray ground color with yellow markings; FEMALE similar
Subspecies 4 described

EUPHYDRYAS CYNTHIA
Cynthia's Fritillary

Size	**Zone**	**Status**
1³/₄ in • 40mm	3	Not protected

Habitat & Ecology alpine marshes
Other Characteristics BELOW base of wings white with yellow spots; FEMALE duller; above, lacks white markings
Subspecies 2 described

♀

EUPHYDRYAS PHAETON
Baltimore

Size | **Zone** | **Status**
2⁵/₈ in • 65mm | 1 | Not protected

Habitat & Ecology localized in moist meadows with *Chelone gabra* (turtlehead); larval food plant: *Plantago lanceolata* (English plantain), *Agalinis* (false foxglove)
Other Characteristics BELOW paler; base of wings suffused with red; FEMALE larger; wings rounder
Subspecies 2 described

EUPHYDRYAS GILLETTI
Gillette's Checkerspot, Yellowstone Checkerspot

Size | **Zone** | **Status**
1⁷/₈ in • 45mm | 1 | P (Yellowstone National Park)

Habitat & Ecology moist mountain meadows; larval food plant: *Lonicera involucrata* (twinberry)
Other Characteristics BELOW muted coloration; MALE smaller
Subspecies none described

EUPLOEA BATESI

Size | **Zone** | **Status**
4 in • 100mm | 6 | Not protected

Habitat & Ecology rain forest clearings and edges, secondary forests to 1,200 ft (360m)
Other Characteristics BELOW paler; FEMALE darker
Subspecies 6 described

Genus Euploea This
very large genus within the Danainae is often called the Crows. Some species are very dark and drab whereas others are colorful. These butterflies are found in a variety of habitats and are associated with the *Asclepiadaceae* (hoya), *Apocynaceae* (dogbane family), and *Moraceae* (mulberry family) as larvae. Thus they are distasteful to predators. The males are noted for their strong pheromone odors.

EUPLOEA CORE GODARTII
Common Indian Crow,
Australian Crow, Oleander Butterfly

Size	Zone	Status
3⁷/₈ in • 95mm	5•6	Not protected

Habitat & Ecology variable, from rain forest to open woods
Other Characteristics BELOW 3 bluish spots near end forewing cell; FEMALE lacks androconial patch
Subspecies 15 described

EUPLOEA CALLITHOE

Size	Zone	Status
4³/₈ in • 110mm	6	Not protected

Habitat & Ecology variable, including mangrove swamps and marginal secondary forests; larval food plants: *Cerbera floribunda* (cassowary plum) and ornamental *Plumeria* (frangipani)
Other Characteristics BELOW paler rust hindwing; FEMALE duller; lacks androconial patch
Subspecies 5 described

EUPLOEA EURIANASSA

Size	Zone	Status
3⁵/₈ in • 90mm	6	Not protected

Habitat & Ecology rain forest and forest edges; flower visitors; larval food plant: *Ichnocarpus frutescens* (member of the dogbane family)
Other Characteristics BELOW paler; bluish white spots near forewing cell; FEMALE with rounder wings; lacks androconial patch
Subspecies none described

EUPLOEA MULCIBER
Striped Blue Crow

Size	Zone	Status
4 in • 100mm	5	Not protected

Habitat & Ecology common on paths and rain forest clearings; larval food plants: *Nerium* (nerium), *Aristolochia* (birthwort), *Ficus* (fig)
Other Characteristics BELOW 4 bluish white spots on hindwing; FEMALE lacks androconial patch
Subspecies more than 25 described

EUPLOEA PHAENARETA

Size	Zone	Status
4³/₈ in • 110mm	5•6	Not protected

Habitat & Ecology variable, rain forests; ornamental trees; town gardens; avid flower visitor; flies slowly
Other Characteristics BELOW pale lavender median spotband on forewing; FEMALE wings more square cut; lacks androconia
Subspecies more than 20 described

EUPLOEA TULLIOLUS KOXINGA
The Dwarf Crow

Size	Zone	Status
2⁵/₈ in • 65mm	5	Not protected

Habitat & Ecology deep rain forest clearings and paths at moderate elevations; larval food plant: *Moraceae* (mulberry family)
Other Characteristics BELOW dull brownish black; FEMALE with the posterior forewing margin straight; lacks the androconial patch
Subspecies more than 20 described

Genus Euthalia

These large, powerful fliers are widely distributed throughout Asia and Indonesia.

EUTHALIA AEETES

Size	Zone	Status
3³/₈ in • 85mm	6	Not protected

Habitat & Ecology forest dweller
Other Characteristics BELOW blue spots in cell on both wings; FEMALE paler
Subspecies 3 described

Genus Fabriciana

These large fritillaries are widely distributed in Eurasia and are closely related to *Argynnis*.

FABRICIANA NIOBE
Niobe Fritillary

Size	Zone	Status
2³/₈ in • 60mm	3	Not protected

Habitat & Ecology moist meadows, open woodlands, and pastures to treeline; larval food plant: *Viola* (violet)
Other Characteristics BELOW silver spots on hindwing; FEMALE darker; markings heavier
Subspecies only 2 described

Genus Geitoneura

This is an Australian genus of large satyrine butterflies.

GEITONEURA KLUGI

Size	Zone	Status
2 in • 50mm	6	Not protected

Habitat & Ecology open grasslands; open woods
Other Characteristics BELOW 3 subapical white spots on forewing; FEMALE paler; larger markings
Subspecies 3 described

Genus Gyrocheilus

This genus is composed of a single species and is known from the southwestern deserts of the United States and Mexico.

GYROCHEILUS PATROBUS TRITONIA
Red-bordered Brown, Red-rim Satyr

Size	Zone	Status
3 in • 75mm	1•2	Not protected

Habitat & Ecology open pine forests; little known about life history
Other Characteristics ABOVE 3 subapical white spots on forewing; dull rust brown along margin; FEMALE paler
Subspecies 2 described

Genus Hamadryas

This genus, of more than 12 species, is known as the Crackers or Calico butterflies because of the "cracking" sound they make as they fly along paths or rest upside down on trees.

HAMADRYAS ATLANTIS
Dusty Cracker

Size	Zone	Status
2⁷/₈ in • 70mm	2	Not protected

Habitat & Ecology dry washes and canyons with mesic tropical forest nearby
Other Characteristics ABOVE black mottled with greenish red; FEMALE larger with a white subapical bar on forewing
Subspecies 2 described

HAMADRYAS AMPHINOME
Red Cracker

Size	Zone	Status
1⁷/₈ in • 45mm	2	Not protected

Habitat & Ecology open woodlands; secondary and mesic tropical rain forest
Other Characteristics BELOW brick red at base; bluish white marginal spots on both wings; FEMALE below, hindwing much duller
Subspecies 4 described

HAMADRYAS FORNAX
Yellow Cracker, Yellow-skirted Cracker

Size	**Zone**	**Status**
1³/₄ in • 40mm	1•2	Not protected

Habitat & Ecology moist forests
Other Characteristics ABOVE marbled shades of gray, dull blue and brown; two black ocelli on submargin; FEMALE rounder forewings
Subspecies 2 described

HAMADRYAS LAODAMIA

Size	**Zone**	**Status**
3 in • 75mm	2	Not protected

Habitat & Ecology clearings in lowland tropical rain forest and along edges; both sexes fly in canopy; larval food plant: *Dalleschampia* (member of nettle family)
Other Characteristics BELOW 3 red spots at base; FEMALE broad white transverse forewing band
Subspecies 4 described

Genus Hamanumida

This genus contains the Guineafowl, so-called because of its coloration. It is widely spread throughout Africa and extends to Saudi Arabia.

HAMANUMIDA DAEDALUS
Guineafowl

Size	**Zone**	**Status**
2³/₈ in • 60mm	4	Not protected

Habitat & Ecology rocky areas of savannah or scrub forest; larval food plant: *Combretaceae* (Jerusalem thorn)
Other Characteristics BELOW rust orange ground color; FEMALE similar
Subspecies none described

Genus Heliconius

Commonly called the Longwings because of their long, narrow forewings, these butterflies are protected from predation by toxic compounds taken from *Passiflora* (passion flower) as larvae. They are slow fliers, but are gregarious and roost communally in trees or bushes.

HELICONIUS ANTIOCHUS

Size	**Zone**	**Status**
3 in • 75mm	2	Not protected

Habitat & Ecology rain forest edges; clearings
Other Characteristics BELOW rust streak along subcosta on hindwing; FEMALE similar
Subspecies none described

HELICONIUS BURNEYI

Size	**Zone**	**Status**
3⅝ in • 90mm	2	Not protected

Habitat & Ecology forest clearings and open woodland
Other Characteristics BELOW similar; FEMALE less vivid in coloration
Subspecies none thus far but there are a number of forms described

HELICONIUS CHARITONIUS TUCKERI
Zebra

Size	**Zone**	**Status**
3⅞ in • 95mm	1•2	Not protected

Habitat open paths in woodland and forests; larval food plant: *Passiflora* (passion flower)
Other Characteristics BELOW red dots on both wings; MALE similar
Subspecies 7 described

♀

HELICONIUS ERATO
The Small Postman

Size	Zone	Status
3⁷/₈ in • 95mm	1•2	Not protected

Habitat & Ecology pastures; open forests; woodlands; disturbed habitats
Other Characteristics both sexes polymorphic; BELOW similar; salmon markings reduced; FEMALE similar
Subspecies none described but numerous forms

HELICONIUS HERMATHENA

Size	Zone	Status
3³/₈ in • 85mm	2	Not protected

Habitat & Ecology mesic scrub; forest edges
Other Characteristics BELOW yellow hindwing margins enlarged; FEMALE similar
Subspecies none described

HELICONIUS MELPOMENE ARMARYLLIS
Postman

Size	Zone	Status
3³/₈ in • 85mm	2	Not protected

Habitat & Ecology open tropical forests; larval food plant: *Passiflora* (passion flower
Other Characteristics highly polymorphic; BELOW similar; FEMALE lacks the pale androconial area on the hindwing anterior margin
Subspecies numerous described but of doubtful validity

HELICONIUS METHARME

Size	**Zone**	**Status**
3¹/₈ in • 80mm	2	Not protected

Habitat & Ecology open clearings; rain forest paths
Other Characteristics BELOW pale yellow streak on forewing; FEMALE lacks androconial patch
Subspecies none described

Genus Hestina

This is a small genus of the "siren butterflies" with rather pronounced indentation of the forewing.

Genus Heteronympha

This is a genus of Australian browns that resemble members of the European *Lasiommata*.

HESTINA NAMA
Circe

Size	**Zone**	**Status**
4¹/₈ in • 105mm	5	Not protected

Habitat & Ecology mesic tropical forests
Other Characteristics BELOW reddish brown areas lighter; FEMALE similar
Subspecies 6 described

HETERONYMPHA MEROPE

Size	**Zone**	**Status**
2⁵/₈ in • 65mm	6	Not protected

Habitat & Ecology grasslands
Other Characteristics ABOVE tawny with black brown toward forewing apex and brown along hindwing margin; FEMALE paler; above, browner
Subspecies 3 described

Genus Hipparchia

This is a group of agile browns from Europe and Asia. Their cryptic coloration affords them protection in the similarly colored dry montane terrain.

♀

HIPPARCHIA SEMELE
Grayling

Size	Zone	Status
2 in • 50mm	3	Not protected

Habitat & Ecology flies in lowland and open montane heaths; larval food plants: *Poaceae* (grass family)
Other Characteristics BELOW similar; MALE darker; lacks the defined forewing median band
Subspecies 2 described

HIPPARCHIA FAGI
Woodland Grayling

Size	Zone	Status
3 in • 75mm	3	Not protected

Habitat & Ecology variable open, mesic woodlands to 3,000 ft (1,000m); larval food plant: *Poaceae* (grass family)
Other Characteristics BELOW similar with white overscaling; FEMALE white markings enlarged
Subspecies a few described

Genus Hyaliris

This is a neotropical genus of clear-winged butterflies.

HYALIRIS IMAGUNCULA

Size	Zone	Status
3 in • 75mm	2	Not protected

Habitat & Ecology rain forest paths and clearings; larval food plants: *Solanaceae* (nightshade family)
Other Characteristics BELOW similar; FEMALE similar
Subspecies none described

Genus Hypna

This highly variable, neotropical genus has unusual angular wing margins.

HYPNA CLYTEMNESTRA

Size	Zone	Status
4 in • 100mm	2	Not protected

Habitat & Ecology low to mesic tropical forests
Other Characteristics BELOW silver at base and along costa of both wings; FEMALE similar
Subspecies 7 described

Genus Hypolimnas

This genus has remarkable sexual dimorphism. These butterflies are also involved in mimicry complexes with various danaids from Asia, Indonesia, and Australia.

▲
♀

HYPOLIMNAS BOLINA
Common Eggfly, Great Eggfly

Size	Zone	Status
4³/₈ in • 110mm	4•5•6	Not protected

Habitat & Ecology variable, secondary forests; rain forests; forest edges and gardens; strong fliers; somewhat migratory; larval food plants include: *Acanthaceae* (acanthia family)
Other Characteristics ABOVE variable from nearly white, to bands with black and blue; MALE black with bluish white spots at cell ends on both wings; white apical spot
Subspecies more than 12 described and a number of female forms

HYPOLIMNAS ALIMENA
Blue-banded Eggfly

Size	Zone	Status
2⁵/₈ in • 65mm	6	Not protected

Habitat & Ecology open areas in moist forests; secondary forests; shrubby areas
Other Characteristics BELOW similar; FEMALE very different; broad white subapical band; often with brown on the outer hindwing margin
Subspecies more than 20 described

HYPOLIMNAS DIOMEA

Size	Zone	Status
4¹/₂ in • 115mm	5•6	Not protected

Habitat & Ecology rain forest in Sulawesi
Other Characteristics BELOW similar; MALE black with bluish purple bands
Subspecies 2 described

♀

HYPOLIMNAS MISIPPUS
Diadem Butterfly,
Five-continent Butterfly, Danaid Eggfly

Size	Zone	Status
2⁵/₈ in • 65mm	1•2•4•5•6	Not protected

Habitat & Ecology strong flier in open woodlands and clearings
Other Characteristics females some form of danaid mimic; BELOW similar; with additional black spot; MALE black with white blue-edged spots in wing cells; white subapical patch
Subspecies 5 described

♀

HYPOLIMNAS PANDARUS PANDORA

Size	Zone	Status
4⁷/₈ in • 120mm	5•6	Not protected

Habitat & Ecology open clearings; woodland edges; avid flower visitor
Other Characteristics BELOW similar; FEMALE row of submarginal white forewing spots
Subspecies 3 described

Genus Hyponephele

This genus of browns from Europe and Asia is quite dimorphic and is associated with various grasses as larvae.

HYPONEPHELE LYCAON
Dusky Meadow Brown

Size	Zone	Status
2 in • 50mm	3	P (locally only)

Habitat & Ecology mesic lowland rocky fields; larval food plants: *Poaceae* (grasses)
Other Characteristics ABOVE similar; FEMALE above, more orange above, especially on hindwing and forewing androconial patches
Subspecies 3 described

Genus Hyposcada

This is a neotropical genus of rather long-winged butterflies. They are associated with the *Solanaceae* (nightshade family) as larvae, and this affords them protection from predators as adults.

HYPOSCADA KEZIA

Size	Zone	Status
2³/₈ in • 60mm	2	Not protected

Habitat & Ecology open paths and edges of dense tropical rain forest
Other Characteristics BELOW similar; FEMALE similar
Subspecies 2 described

Genus Idea

The species of this genus are all different shades and combinations of black, gray, and white. These butterflies fly rather weakly in dense forests of Asia, Indonesia, and Australia.

IDEA DURVILLEI

Size	Zone	Status
5⁷/₈ in • 145mm	6	Not protected

Habitat & Ecology dense rain forest
Other Characteristics BELOW similar; MALE similar
Subspecies 6 described

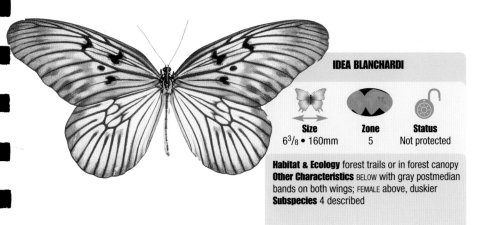

IDEA BLANCHARDI

Size
6³/₈ • 160mm

Zone
5

Status
Not protected

Habitat & Ecology forest trails or in forest canopy
Other Characteristics BELOW with gray postmedian bands on both wings; FEMALE above, duskier
Subspecies 4 described

IDEA JASONIA

Size
5 in • 125mm

Zone
5

Status
Not protected

Habitat & Ecology open paths; clearings; edges of tropical rain forest
Other Characteristics BELOW similar; FEMALE similar
Subspecies none described

IDEA TAMBUSISIANA

Size
7 in • 175mm

Zone
6

Status
Rare

Habitat & Ecology primary rain forest
Other Characteristics BELOW similar; FEMALE ground color yellowish
Subspecies none described

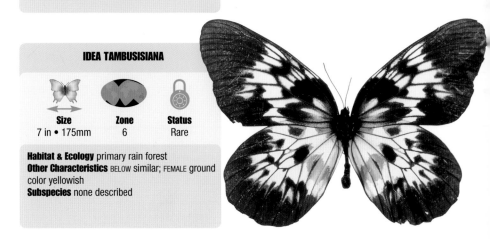

Genus Ideopsis

This is an Asian genus related to *Idea*, and composed of 7 species.

IDEOPSIS HEWITSONI

Size	Zone	Status
4 in • 100mm	6	Not protected

Habitat & Ecology variable; marginal secondary forests; open areas; shrubs
Other Characteristics BELOW similar; MALE has more prominent darker markings
Subspecies none described

Genus Inachis

This genus is composed of a single species, and is closely related to European *Nymphalis*.

INACHIS IO
Peacock

Size	Zone	Status
2³/₈ in • 60mm	3	Not protected

Habitat & Ecology flies in disturbed and open wooded areas with flowers
Other Characteristics BELOW mottled brown forewing ocellus; postmedian line on hindwing; FEMALE less brightly colored
Subspecies none described

Genus Junonia

These are the so-called Buckeyes or Inspectors and are widely distributed.

JUNONIA ALMANA
Peacock Pansy

Size	Zone	Status
2⁵/₈ in • 65mm	5	Not protected

Habitat & Ecology various, open woods; gardens; larval food plant: *Mimosa pudica* (touch-me-not)
Other Characteristics BELOW similar; FEMALE similar
Subspecies 3 described

JUNONIA COENIA
Buckeye

Size	Zone	Status
2⁵/₈ in • 65mm	1•2	Not protected

Habitat & Ecology mesic tropical forests; larval food plant: *Plantago major* (plantain)
Other Characteristics BELOW green lines on hindwing; FEMALE less sharp forewing tips
Subspecies 3 described

JUNONIA EVARETE ZONALIS
West Indian Buckeye, Florida Buckeye, Tropical Buckeye

Size	Zone	Status
2⁵/₈ in • 65mm	1•2	Not protected

Habitat & Ecology open tropical woodlands; larval food plant: *Avicennia germinans* (black mangrove)
Other Characteristics BELOW forewing overscaled with rust; FEMALE more rectangular forewings
Subspecies 4 described

JUNONIA HIERTA
Yellow Pansy

Size	Zone	Status
2¹/₈ in • 55mm	4•5	Not protected

Habitat & Ecology savannah; grassy slopes; males territorial along paths; larval food plants: various *Acanthaceae* (acanthia family)
Other Characteristics BELOW similar; FEMALE paler
Subspecies none described

♀

JUNONIA OCTAVIA
Gaudy Commodore

Size	Zone	Status
2⁷/₈ in • 70mm	4	Not protected

Habitat & Ecology savannah woodlands; larval food plants: various *Labiatae* (mint family)
Other Characteristics 2 color morphs in both sexes: orange (dry season) and blue (wet season); BELOW similar; MALE similar
Subspecies none described

JUNONIA ORITHYA MADAGASCARENSIS
Eyed Pansy, Blue Pansy

Size	Zone	Status
2¹/₈ in • 55mm	4•5•6	Not protected

Habitat & Ecology open woodlands; fields; scrub areas; males very territorial
Other Characteristics BELOW tan ground color; 3 orange bars in forewing cell FEMALE similar
Subspecies more than 15 described

Genus Kallima
These butterflies are commonly called the "leaf butterflies" because of their cryptic coloration below.

KALLIMA HORSFIELDI PHILARCHUS
Blue Oak Leaf Butterfly

Size	Zone	Status
4³/₈ in • 110mm	5	Not protected

Habitat open woodlands
Other Characteristics BELOW olive brown suffused with lavender; rust postmedian line on hindwing; FEMALE similar
Subspecies 2 generally described

KALLIMA PARALEKTA
The Indian Leaf

Size	Zone	Status
4 in • 100mm	5	Not protected

Habitat & Ecology dense forests up to moderate elevations
Other Characteristics BELOW olive brown shaded with lavender; FEMALE above, brown, with white bar across forewing and a white subapical spot
Subspecies 3 described

KALLIMA RUMIA
African Leaf Butterfly

Size	Zone	Status
3¹/₈ in • 80mm	4	Not protected

Habitat & Ecology open woodland and forests
Other Characteristics BELOW golden brown ground color; FEMALE dark brown; pale blue bar across forewing and whitish hindwing patch
Subspecies 3 described

Genus Lasiommata

These browns appear similar to *Parage* and are widely distributed through Europe and and Asia. Their larval food plants include *Poaceae* (grasses).

LASIOMMATA MAERA
Large Wall Brown

Size	Zone	Status
5¹/₈ in • 130mm	3	Not protected

Habitat & Ecology rocky montane meadows and slopes up to 6,000 ft (2,000m)
Other Characteristics ABOVE brown; dark brown androconia below cell; FEMALE more orange above with large apical forewing ocellus; generally 2 submarginal ocelli on hindwing
Subspecies 4 described

LASIOMMATA MEGERA
The Wall

Size	Zone	Status
2 in • 50mm	3	Not protected

Habitat & Ecology open country; usually rough and rocky terrain
Other Characteristics BELOW paler; FEMALE lacks the enlarged forewing androconial patch
Subspecies 2 described

Genus Lethe

This is a widely distributed genus
of medium-sized brown butterflies.

♀

▲
♀

LETHE APPALACHIA

*Appalachian Brown,
Woods Eyed Brown*

Size	Zone	Status
2 in • 50mm	1	P (Limited)

Habitat & Ecology moist marshes and meadows; larval food plant: *Carex* (sedge family)
Other Characteristics ABOVE brown, with forewing ocelli; MALE smaller ocelli
Subspecies 2 described

LETHE DARENA SUMATRENSIS

Size	Zone	Status
3¹/₈ in • 80mm	5	Not protected

Habitat & Ecology dense montane woodlands
Other Characteristics ABOVE dark brown with the prominent forewing yellow band; blackish brown submarginal ocelli on hindwing above;
MALE similar
Subspecies more than 3 described

LETHE EUROPA
Bamboo Tree-brown

Size	Zone	Status
3 in • 75mm	5	Not protected

Habitat & Ecology anywhere with the larval food plant: *Poaceae* (bamboo)
Other Characteristics ABOVE uniform dark brown; broad whitish forewing band; FEMALE similar
Subspecies more than 12 described

LETHE PORTLANDIA
Southern Pearly Eye

Size	Zone	Status
2⁷/₈ in • 70mm	1	Not protected

Habitat & Ecology moist woodlands; frequently perch on trees
Other Characteristics ABOVE olive green with blind ocelli (without eyes) repeated; FEMALE rounder forewing apex
Subspecies 3 described

Genus Lexias

This is a group of large forest butterflies found in the Indian and Australian regions.

LEXIAS AEROPA

Size	Zone	Status
3¹/₈ in • 80mm	6	Not protected

Habitat & Ecology clearings of dense forests, secondary forests near streams; infrequently encountered; larval food plant: *Calophyllum* (calaba)
Other Characteristics BELOW similar; FEMALE brown with a white forewing median spot band and a white hindwing postdiscal spot band
Subspecies 8 described

LEXIAS DIRTEA

Size	Zone	Status
4¹/₂ in • 115mm	5	Not protected

Habitat & Ecology open forest up to 1,500 ft (500m); attracted to fermenting fruit; larval food plant: *Garcinia laterifolia* (garcinia)
Other Characteristics BELOW ground color golden brown; lighter markings; FEMALE above, brown; peppered with yellow spots over both wings
Subspecies none described

LEXIAS SATRAPES ORNATA

Size	Zone	Status
4¹/₂ in • 115mm	5	Not protected

Habitat & Ecology deep dense forests
Other Characteristics BELOW hindwing red brown; FEMALE larger; above, larger white spots
Subspecies 5 described

Genus Libythea

These are the snout butterflies which have elongated palpi that protrude forward. The larvae feed on *Celtis* (hackberry).

LIBYTHEA CELTIS
Nettle-tree Butterfly

Size	Zone	Status
1³/₄ in • 40mm	3•5	Not protected

Habitat & Ecology open highland woodland, fields; larval food plant: *Celtis australis* (nettle tree)
Other Characteristics BELOW paler; hindwing dark brown, mottled with white; FEMALE 2 dark brown spots along forewing apex
Subspecies 4 described

LIBYTHEA GEOFFROYI MAENIA
Beak Butterfly

Size	**Zone**	**Status**
2³/₈ in • 60mm	5•6	Not protected

Habitat & Ecology open woodland; secondary forests along streams; larval food plants: include *Lauraceae* (laurel family)
Other Characteristics BELOW lacks blue; hindwing darker; FEMALE brown with orange or white forewing spots and orange hindwing bar
Subspecies more than 15 described

LIBYTHEA LABDACA
African Snout

Size	**Zone**	**Status**
2 in • 50mm	4	Not protected

Habitat & Ecology strongly migratory; found in savannah; edges of woods; coastal forests; larval food plant: *Celtis* (hackberry)
Other Characteristics BELOW similar; FEMALE up to 4 in (100mm) smaller than males
Subspecies 3 described

Genus Libytheana
Composed of 4 species, these snout butterflies are restricted to the western hemisphere.

LIBYTHEANA BACHMANNI
Snout Butterfly

Size	**Zone**	**Status**
2 in • 50mm	1•2	Not protected

Habitat & Ecology scrub habitats; desert washes; strongly migratory
Other Characteristics BELOW forewing apex overscaled with white; FEMALE similar
Subspecies 2 described

Genus Limenitis This is a rather diverse genus with representatives in Europe, Asia, and Australia. The young larvae hibernate.

LIMENITIS LYMIRE

Size	Zone	Status
3³/₈ in • 85mm	5	Not protected

Habitat & Ecology edges of woods; paths; clearings
Other Characteristics BELOW purplish brown to pale green; veins edged in white; FEMALE similar
Subspecies none described

LIMENITIS POPULI
Poplar Admiral

Size	Zone	Status
3¹/₈ in • 80mm	3	Not protected

Habitat & Ecology open woodlands
Other Characteristics BELOW greenish gray along margins; FEMALE white markings larger
Subspecies more than 8 described

Genus Lopinga This genus of browns are characterized by their prominent rows of eyespots.

LOPINGA ACHINE
Woodland Brown

Size	Zone	Status
2¹/₈ in • 55mm	3	Not protected

Habitat & Ecology shady open woods up to 3,000 ft (1,000m); larval food plant: *Poaceae* (grass family)
Other Characteristics ABOVE similar; FEMALE similar
Subspecies none described

Genus Lycorea

This is a group of longwings which are part of the *Danainae*.

LYCOREA CLEOBAEA ATERGATIS

Size	Zone	Status
4 in • 100mm	2	Not protected

Habitat & Ecology variable, but generally paths and edges of mesic tropical forests; larval food plants: include *Ficus* (fig)
Other Characteristics BELOW darker; FEMALE similar
Subspecies 8 described

Genus Maniola

This is a group of temperate browns that are widely distributed in Europe and Asia and appear to be closely related to some South American species.

MANIOLA JURTINA
Meadow Brown

Size	Zone	Status
2 in • 50mm	3	Not protected

Habitat & Ecology grasslands and meadows; larval food plants: *Poaceae* (grass family)
Other Characteristics ABOVE similar; FEMALE dull brown with single forewing ocellus larger, and enclosed in an orange band
Subspecies 10 described

Genus Marpesia

This is a genus of tailed butterflies that are somewhat migratory, and strong angular fliers.

MARPESIA PETREUS
*Ruddy Dagger Wing, Red Dagger Wing,
Southern Dagger Tail*

Size	Zone	Status
3$^{1}/_{8}$ in • 80mm	1•2	Not protected

Habitat & Ecology variable, open marshy woodlands; mesic tropical deciduous forests; larval food plant: *Ficus* (fig)
Other Characteristics BELOW cryptic coloration; mottled purple; FEMALE above, duller
Subspecies 2 described

Genus Megisto

This is a small genus of rather small browns that are very common where they occur.

MEGISTO RUBRICATA
Red Satyr

Size	Zone	Status
2 in • 50mm	1•2	Not protected

Habitat & Ecology grasslands; open fields; woodlands in canyons; larval food plants: *Poaceae* (grass family)
Other Characteristics BELOW rust postmedian lines on both wings; FEMALE duller
Subspecies 4 described

Genus Melanargia

The characteristic white ground color of this genus of browns gives them their common name—Marbled Whites.

♀

MELANARGIA GALATHEA
Marbled White

Size	Zone	Status
2 in • 50mm	3	Not protected

Habitat & Ecology meadows; open woodlands; larval food plant: *Poaceae* (grass family)
Other Characteristics BELOW similar; MALE more crisply marked; duskier white markings
Subspecies 5 described

Genus Melanitis

The 25 species of this genus are all medium-sized browns, some of which are migratory.

▲

MELANITIS AMABILIS

Size	Zone	Status
3¹/₈ in • 80mm	6	Not protected

Habitat & Ecology rain forest; secondary vegetation; larval food plant: *Poaceae* (grass family)
Other Characteristics ABOVE dark brown with yellow forewing bar; FEMALE dull dark brown with a white bar across forewing
Subspecies 5 described

MELANITIS LEDA
Evening Brown,
Common Evening Brown

Size	Zone	Status
3³/₈ in • 85mm	4•5•6	Not protected

Habitat & Ecology woods; flies at dusk; migratory; larval food plant: *Poaceae* (grass family)
Other Characteristics BELOW dark brown overscaled with white; FEMALE forewing spots larger
Subspecies more than 12 described

Genus Melinaea

This is a genus of longwings within the *Ithomiinae*. They are associated with *Solanaceae* (nightshade family) as larvae and are distasteful to predators.

♀

MELINAEA LILIS MESSATIS

Size	Zone	Status
3 in • 75mm	2	Not protected

Habitat & Ecology variable, paths and clearings in rain and tropical dry forests; disturbed habitats; larval food plants: *Solanaceae* (nightshade family)
Other Characteristics BELOW paler; FEMALE similar
Subspecies 20 described

MELINAEA ZANEKA

Size	Zone	Status
3⁷/₈ in • 95mm	2	Not protected

Habitat & Ecology clearings in rain forests
Other Characteristics BELOW duller with white marginal spots; MALE similar
Subspecies none described

Genus Melitaea
This is a genus of smaller fritillaries restricted to the northern hemisphere.

MELITAEA CINXIA
Glanville Fritillary

Size	Zone	Status
1³/₄ in • 40mm	3	Not protected

Habitat & Ecology flowery meadows; heaths
Other Characteristics BELOW forewing yellow at apex; FEMALE larger; ground color more tan
Subspecies more than 12 described

MELITAEA DIDYMA
Spotted Fritillary

Size	Zone	Status
1⁷/₈ in • 45mm	3	Not protected

Habitat & Ecology meadow dweller; larval food plant: *Plantago major* (plantain)
Other Characteristics BELOW basal orange band on hindwing; FEMALE larger; above, duller orange
Subspecies more than 2 described

Genus Memphis
Originally incorporated into *Anaea*, this genus is quite distinct. It is composed of 40 species with cryptic coloration below.

MEMPHIS AUREOLA

Size	Zone	Status
2⁵/₈ in • 65mm	2	Not protected

Habitat & Ecology edges and paths of lowland rain forests, up to 2,500 ft (750m); males are pugnacious along trails
Other Characteristics BELOW hindwing mottled; FEMALE tailed; duller in coloration
Subspecies none described

Genus Minois

These are medium-sized browns or satyrids.

MINOIS DRYAS
The Dryad

Size	Zone	Status
2⁵/₈ in • 65mm	3	Not protected

Habitat & Ecology montane meadows
Other Characteristics BELOW paler;
FEMALE paler; above, slightly larger eyespots
Subspecies few of doubtful validity described

Genus Morpho

With a variety of wing colorations, brown, pearl white, and various shades of matte or iridescent aquamarine blue, these butterflies are favorites of collectors. Adults are attracted to fermenting fruit, or sap flows.

MORPHO AEGA
Morpho

Size	Zone	Status
3⁵/₈ in • 90mm	2	P (Limited)

Habitat & Ecology common along trails in open tropical forests
Other Characteristics BELOW tan ground color; ocelli on both wings; FEMALE brownish and yellow, often with blue highlights, and a bar at the cell end on the forewing
Subspecies none described

MORPHO CYPRIS
Morpho

Size	Zone	Status
3⁵/₈ in • 90mm	2	P (Limited)

Habitat & Ecology rain forest; females fly in the canopy and are rarely seen
Other Characteristics BELOW brown and white; 2 median bands and spot near forewing end cell; FEMALE dimorphic
Subspecies none described

MORPHO PELEIDES CORYDON
Morpho

Size	**Zone**	**Status**
6¹/₈ in • 155mm	2	Not protected

Habitat & Ecology clearings and paths in rain forest
Other Characteristics BELOW black; paler on forewing; FEMALE similar with broader black
Subspecies more than 11 described

Genus Mycalesis
This genus is composed of 60 species, which are widely distributed in Asia and Australia.

MYCALESIS DUPONCHELI

Size	**Zone**	**Status**
2³/₈ in • 60mm	6	Not protected

Habitat & Ecology deep forests; flies at dusk
Other Characteristics ABOVE deep rich brown and yellow is replaced with orange; FEMALE paler; yellow along lateral margin of hindwing and along posterior margin of forewing
Subspecies 9 described

MYCALESIS MESSENE

Size	**Zone**	**Status**
2 in • 50mm	6	Not protected

Habitat & Ecology deep forest; woods
Other Characteristics BELOW 2 ocelli on forewing; FEMALE larger; paler; ocelli enlarged
Subspecies none described

Genus Mynes

Composed of 8 species, this genus resembles some pierids and is quite variable in coloration.

MYNES GEOFFROYI
White Nymph

Size	Zone	Status
2⁷/₈ in • 70mm	6	Not protected

Habitat & Ecology woodlands
Other Characteristics ABOVE creamy yellow; FEMALE duller; more gray and black with a dark border on forewing; broader border on hindwing
Subspecies 6 described

Genus Neominois

This genus is composed of a single species.

Genus Neope

These large satyrid butterflies are endemic to southeast Asia.

NEOPE BHADRA

Size	Zone	Status
3⁵/₈ in • 90mm	5	Not protected

Habitat & Ecology generally rain forest
Other Characteristics BELOW hindwing overscaled with light lavender; FEMALE larger; paler
Subspecies none described

NEOMINOIS RIDINGSII
Riding's Satyr

Size	Zone	Status
1⁷/₈ in • 50mm	1	Not protected

Habitat & Ecology grasslands and plains; larval food plant: *Poaceae* (grass family)
Other Characteristics BELOW darker postmedian band on hindwing; FEMALE paler
Subspecies 4 described

Nymphalidae

Genus Neorina
Endemic to Southeast Asia and Indonesia, these butterflies are rather large, with long sensory hairs on the upper surface.

NEORINA HILDA

Size	Zone	Status
3⁷/₈ in • 85mm	5	Not protected

Habitat & Ecology along paths and edges of woodlands and forests; invades open areas
Other Characteristics ABOVE brown with the broad pale yellow orange bar across forewing; FEMALE larger
Subspecies more than 2 described

NEORINA LOWII
Malayan Owl

Size	Zone	Status
4¹/₈ in • 105mm	5	Not protected

Habitat & Ecology open woodlands
Other Characteristics ABOVE dark brown with broad yellow band near forewing tornus and on hindwing apex; FEMALE similar
Subspecies 5 described

Genus Neptis
This large genus is composed of 50 species found in Europe, Asia, Africa, and Australia. They are commonly called the Sailors and are denizens of savannahs and open woods.

NEPTIS MAHENDRA

Size	Zone	Status
2³/₈ in • 60mm	5	Not protected

Habitat & Ecology open woods; forest edges
Other Characteristics BELOW dull rust band on both wings; MALE similar, but inner margin of hindwing enlarged
Subspecies 5 described

NEPTIS SACLAVA
Small Spotted Sailor

Size	Zone	Status
1⁷/₈ in • 45mm	4	Not protected

Habitat & Ecology open savannah
Other Characteristics ABOVE blackish brown; brown submarginal band; FEMALE similar
Subspecies 2 described

NEPTIS ZAIDA MANIPURENSIS

Size	Zone	Status
2⁵/₈ in • 65mm	5	Not protected

Habitat & Ecology open woodlands; savannah
Other Characteristics BELOW dull rust brown; MALE similar
Subspecies 6 described

Genus Neurosigma

This genus has a single species and is restricted to India only.

NEUROSIGMA SIVA

Size	Zone	Status
2⁷/₈ in • 70mm	5	Not protected

Habitat & Ecology woodlands
Other Characteristics BELOW paler; duller; FEMALE similar but ground color basically cream with orange on base of wings and heavier dark markings
Subspecies 2 described

Genus Nymphalis
This small genus of strong fliers is restricted to the northern temperate zone. They are frequently encountered in a variety of habitats.

NYMPHALIS ANTIOPA
Mourning Cloak, Camberwell Beauty

Size	Zone	Status
2⅝ in • 65mm	1•2•3	Not protected

Habitat & Ecology open woodlands; forest edges; along streams; adults hibernate
Other Characteristics BELOW similar; MALE similar
Subspecies 3 described

NYMPHALIS VAU-ALBUM
Compton Tortoiseshell, False Comma

Size	Zone	Status
2⅝ in • 65mm	1•3	Not protected

Habitat & Ecology temperate woodland
Other Characteristics BELOW blue submarginal band on both wings; FEMALE similar
Subspecies nearly 20 described

Genus Oeneis
Commonly called the Arctics, these butterflies with their cryptic coloration are restricted to arctic areas of North America, Europe, and Asia. Many species are biennial—a species will hibernate one year as larvae and the following year as pupae.

OENEIS BORE MCKINLEYENSIS
Arctic Grayling

Size	Zone	Status
2 in • 50mm	1•3	Not protected

Habitat & Ecology tundra
Other Characteristics ABOVE dull brown; darker margins; MALE smaller; forewing apex sharper
Subspecies 10 described

OENEIS CHRYXSUS
Chryxsus Arctic, Brown Arctic

Size	**Zone**	**Status**
2 in • 50mm	1	Not protected

Habitat & Ecology open taiga and tundra
Other Characteristics BELOW mottled forewing
with gray at apex; FEMALE above, paler
Subspecies 6 described

OENEIS JUTTA
Forest Arctic, Baltic Grayling,
Jutta Arctic

Size	**Zone**	**Status**
2¹/₈ in • 55mm	1•3	Not protected

Habitat & Ecology black spruce and tamarack
bogs; larval food plant: *Carex* (sedge family)
Other Characteristics ABOVE dull brown; orange
patches or circles around ocelli; androconial
patch below cell; FEMALE ocelli enlarged
Subspecies 12 described

OENEIS MACOUNII
Canada Arctic, Macoun's Arctic

Size	**Zone**	**Status**
2¹/₈ in • 55mm	1	Not protected

Habitat & Ecology woodland species; frequently
rests on tree trunks
Other Characteristics BELOW paler forewing;
overscaled with white; FEMALE larger ocelli
Subspecies none described

♀
▲

OENEIS TAYGETE
White-veined Arctic, Labrador Arctic

Size	Zone	Status
2 in • 50mm	1•3	Not protected

Habitat & Ecology moist grassy areas; tundra, above timber line; larval food plants: *Carex* (sedge family)
Other Characteristics ABOVE dark brownish yellow and unmarked; MALE similar
Subspecies 12 described

Genus Olyras

The 3 species of this neotropical genus are characterized by transparent patches on their long wings.

Genus Opsiphanes

This is a genus of crepuscular, satyrid butterflies with rather robust bodies.

OLYRAS PRAESTANS

Size	Zone	Status
3⁷/₈ in • 95mm	2	Not protected

Habitat & Ecology paths and clearings in lowland and montane rain forest
Other Characteristics BELOW similar; FEMALE similar
Subspecies 3 described

OPSIPHANES BATEA

Size	Zone	Status
3⁵/₈ in • 90mm	2	Not protected

Habitat & Ecology rain forest; areas with fruit
Other Characteristics BELOW similar; FEMALE similar but much larger; wings more rounded
Subspecies 3 described

Genus Oreixenica

These are small brown butterflies endemic to Australia.

OREIXENICA KERSHAWI
Kershaw's Brown

Size	Zone	Status
1³/₄ in • 40mm	6	Not protected

Habitat & Ecology grasslands
Other Characteristics BELOW silvery stripes on both wings; FEMALE slightly larger; lacks the androconia
Subspecies none described

Genus Pantoporia

This genus of butterflies is endemic to Asia and Australia. They are very strong fliers.

PANTOPORIA EULIMENE BADOURA

Size	Zone	Status
3³/₈ in • 85mm	5•6	Not protected

Habitat & Ecology slow flying and gliding in secondary forests, where they perch on leaves with the wings flat
Other Characteristics BELOW paler; FEMALE similar
Subspecies 4 described

Genus Parantica

This genus was originally incorporated into *Danaus* but has been shown subsequently to be distinct. These butterflies are variable in coloration and may be pale yellow or various shades of brown.

PARANTICA SCHENKI

Size	Zone	Status
3¹/₈ in • 80mm	6	Not protected

Habitat & Ecology open areas and forests up to 5,000 ft (1,600m)
Other Characteristics BELOW similar; FEMALE more white; yellow reduced to base of both wings
Subspecies 5 described

PARANTICA VITRINA

Size	Zone	Status
3¹/₈ in • 80mm	5	Not protected

Habitat & Ecology open woodlands; forest edges
Other Characteristics BELOW similar; FEMALE larger, with larger, darker markings
Subspecies 2 described

Genus Pararge

This is a group of medium-sized satyrids that occur widely in Asia and Europe. The larval food plants include *Poaceae* (grass family).

PARARGE AEGERIA TIRCIS
Speckled Wood

Size	Zone	Status
1⁷/₈ in • 45mm	3	Not protected

Habitat & Ecology shady woodlands; normally rest on trees and are well camouflaged
Other Characteristics BELOW similar; FEMALE similar
Subspecies 2 described

Genus Parasarpa

This Asian genus includes large butterflies that are related to the European *Apatura*. They are strong fliers, and the males are pugnacious and quite territorial.

PARASARPA DUDU

Size	Zone	Status
4³/₈ in • 110mm	5	Not protected

Habitat & Ecology open woodland; forest edges; larval food plant: *Loniceas* (honeysuckle)
Other Characteristics BELOW reddish brown, overscaled with lavender; FEMALE wings rounder
Subspecies 3 described

Genus Parthenos

These butterflies are noted for their distinctive, strong, gliding flight and are commonly found in Asia, Indonesia, and Australia.

PARTHENOS SYLVIA LILACINUS
Clipper

Size	Zone	Status
3⁵/₈ in • 90mm	5•6	Not protected

Habitat & Ecology open woodland; secondary forest; scrub; along streams; avid flower visitors; larval food plants: include *Tinospora* (member of the moonseed family)
Other Characteristics BELOW similar; FEMALE wings are rounder
Subspecies more than 20 described

Genus Pedaliodes

This is a very large genus of medium-sized satyrid butterflies, that are found in the Andes. They are common where they occur.

PEDALIODES PHAEDRA

Size	Zone	Status
2³/₈ in • 60mm	2	Not protected

Habitat & Ecology highland paramo; larval food plant: *Poaceae* (bamboo)
Other Characteristics ABOVE black; FEMALE above, brownish black with whitish bars on forewing
Subspecies none described

Genus Perisama

This is a rather large neotropical genus of more than 40 species that occur in upland tropical forests and can often be seen at mud puddles. These butterflies are very brightly colored above, with more muted color and patterns below.

PERISAMA VANINKA

Size	Zone	Status
2 in • 50mm	2	Not protected

Habitat & Ecology upland rain forest paths
Other Characteristics ABOVE black with blue green bands; FEMALE similar
Subspecies none described

Genus Phaedyma

This genus is closely related to *Neptis*, but recognized as distinct because of its structural differences.

♀

PHAEDYMA SATINA

Size	Zone	Status
2³/₈ in • 60mm	6	Not protected

Habitat & Ecology edges of woods
Other Characteristics BELOW similar; MALE similar
Subspecies none described

Genus Philaethria

This neotropical genus is divided into 3 species. They are associated with *Passiflora* (passion flower) as larvae.

Genus Phalanta

This genus of commonly encountered fritillary butterflies is restricted to African, Asian, and Australian tropics.

PHALANTA PHALANTHA
Common Leopard

Size	Zone	Status
2⁵/₈ in • 65mm	4•5	Not protected

Habitat & Ecology open savannah and bush; mesic forests; avid flower visitors; larval food plants: *Populus alba* (white poplar), *Salix* (willow)
Other Characteristics BELOW hindwing margin overscaled with lavender; FEMALE larger; paler
Subspecies more than 6 described

PHILAETHRIA DIDO
Scarce Bamboo Page

Size	Zone	Status
4³/₈ in • 110mm	2	Not protected

Habitat & Ecology tropical rain forest up to 4,000 ft (1,250m)
Other Characteristics BELOW green spots often overscaled with gray; FEMALE green less intense
Subspecies 4 described, some of doubtful validity

Genus Phyciodes

Originally this genus contained more than 100 species worldwide, but it has been subdivided and is now restricted to the Americas. Most species are tropical, but some occur in the more temperate areas. These butterflies are characterized by their crescent markings below.

PHYCIODES PHAON
Cayman Crescentspot, Phaon Crescentspot

Size	Zone	Status
2⁵/₈ in • 65mm	1•2	Not protected

Habitat & Ecology scrub country and open meadows; some seasonally variable
Other Characteristics BELOW some gray overscaling; FEMALE larger and somewhat paler
Subspecies none described

PHYCIODES CASTILLA

Size	Zone	Status
1⁷/₈ in • 45mm	2	Not protected

Habitat & Ecology forested areas
Other Characteristics ABOVE similar; FEMALE yellow bar across the forewing; hindwing dark border with orange radiating out from base
Subspecies none described

PHYCIODES TEXANA
Texas Crescent

Size	Zone	Status
1³/₄ in • 40mm	1•2	Not protected

Habitat & Ecology shady open scrub; larval food plant: *Acanthaceae* (acanthia family)
Other Characteristics BELOW orange at base of forewing; FEMALE larger, broader wings
Subspecies 2 described

Genus Physcopedaliodes

This genus is closely related to *Pedaliodes* but is recognized as distinct by its obvious structural differences.

PHYSCOPEDALIODES PHYSCOA

Size	Zone	Status
2⁷/₈ in • 70mm	2	Not protected

Habitat & Ecology paramo species
Other Characteristics ABOVE dark blackish brown; lacks mottled appearance; FEMALE similar
Subspecies none described

Genus Pierella

Closely related to *Cithaerias*, members of this genus are more heavily patterned and fly only in the understory of primary rain forest.

PIERELLA HORTONA MICROMACULATA

Size	Zone	Status
2³/₈ in • 60mm	2	Not protected

Habitat & Ecology flies in the understory (generally near the ground) of primary rain forest
Other Characteristics BELOW tan; darker in basal area; MALE smaller; paler
Subspecies 4 described

Genus Polygonia

These are the Commas and Anglewings, which have distinctive angular wings and are characterized below by the cryptic bark coloration.

POLYGONIA C-ALBUM
Comma

Size	Zone	Status
2 in • 50mm	3	Not protected

Habitat & Ecology open woods; edges of woods; gardens up to 6,000 ft (2,000m); larval food plants: *Urticaceae* (nettles), *Salix* (willow)
Other Characteristics BELOW below, unmarked dark brown; FEMALE similar
Subspecies numerous described

Genus Polygrapha

Butterflies in this striking genus are closely related to *Charaxes* and *Consul* (originally *Anaea*).

POLYGONIA FAUNUS
Green Comma, Faunus Anglewing

Size	Zone	Status
2 in • 50mm	1	Not protected

Habitat & Ecology open woods and edges of woods; larval food plants: include *Salix* (willow), *Betula* (birch), and *Alnus* (alder)
Other Characteristics ABOVE orange with black and blackish brown markings and margins; FEMALE below, darker
Subspecies 4 described

POLYGONIA INTERROGATIONIS
Question Mark

Size	Zone	Status
2³⁄₈ in • 60mm	1	Not protected

Habitat & Ecology open woodlands; scrub brush
Other Characteristics BELOW characteristic silver comma and centered dot on medial hindwing; FEMALE paler
Subspecies none described

POLYGRAPHA CYANEA

Size	Zone	Status
3¹⁄₈ in • 80mm	2	Not protected

Habitat & Ecology rain forest paths and clearings
Other Characteristics BELOW red brown mottled with slivery white; blue spot band on hindwing; FEMALE above, brown, with orange bars across the forewing and orange hindwing submargin
Subspecies none described

Genus Polyura

This genus of powerful flying butterflies is restricted to Asia, Australia, and New Guinea. These butterflies are closely related to the *Charaxes*.

POLYURA EUDAMIPPUS
Great Nawab

Size	Zone	Status
3¹/₈ in • 80mm	5•6	Not protected

Habitat & Ecology secondary rain forest and forest openings; paths and adjacent gardens
Other Characteristics BELOW ground color silvery white; FEMALE larger; wings rounder
Subspecies 7 described

POLYURA GALAXIA

Size	Zone	Status
5¹/₈ in • 130mm	5•6	Not protected

Habitat & Ecology open areas; forest edges
Other Characteristics ABOVE black with yellow patches at base of wings; FEMALE larger
Subspecies more than 6 described

Genus Prepona

These strong-flying butterflies are related to both *Agrias* and *Anaea*. They are normally observed in the canopy of neotropical rain forest and mesic tropical forests.

PREPONA PHERIDAMAS

Size	Zone	Status
3⁷/₈ in • 95mm	2	Not protected

Habitat & Ecology deep rain forests; tropical forest clearings, paths, and edges; larval food plants: *Fabaceae* (legume/pea family)
Other Characteristics ABOVE ground color blackish brown; iridescent blue to bluish green bands across both wings; FEMALE similar
Subspecies none described

PREPONA PRAENESTE BUCKLEYANA

Size
4 in • 100mm

Zone
2

Status
Not protected

Habitat & Ecology paths and edges of primary rain forest
Other Characteristics BELOW rust ground color FEMALE duller; lacks androconial patches
Subspecies 2 described

Genus Pronophila

This is a genus of more than 40 species of large brown butterflies from the neotropics.

PRONOPHILA CORDILLERA

Size
3³/₈ in • 85mm

Zone
2

Status
Not protected

Habitat & Ecology bamboo forests in the paramo
Other Characteristics ABOVE dark brown but paler toward wing margin; FEMALE similar
Subspecies none described

PRONOPHILA THELEBE

Size
3 in • 75mm

Zone
2

Status
Not protected

Habitat & Ecology montane bamboo forests
Other Characteristics ABOVE black with a few white forewing apical markings; FEMALE similar
Subspecies none described

Genus Prothoe

This genus of strong-flying butterflies is composed of 4 species that are endemic to Australia with 1 in Asia.

PROTHOE CALYDONIA
The Glorious Begum

Size	Zone	Status
4⁷/₈ in • 120mm	5	Not protected

Habitat & Ecology clearings and edges in shady forests; seeks out sap flows and fermenting fruit
Other Characteristics ABOVE black with yellow patch; silvery gray on hindwing; FEMALE similar
Subspecies 6 described

Genus Pseudacraea

Members of this African genus are structurally distinct and mimic *Acraea*.

PSEUDACRAEA BOISDUVALI
Trimen's False Acraea

Size	Zone	Status
4 in • 100mm	5	Not protected

Habitat & Ecology paths; edges of woods; brush
Other Characteristics BELOW pale yellow along hindwing margin; FEMALE larger; wings rounder
Subspecies none described

Genus Ptychandra

These are dimorphic butterflies restricted to Asia.

PTYCHANDRA LORQUINI

Size	Zone	Status
2¹/₈ in • 55mm	5	Not protected

Habitat & Ecology forest clearings and edges
Other Characteristics ABOVE vivid blue; veins and margins dark brown; FEMALE brown with white markings on forewing disc and toward the apex of hindwing; several eyespots on hindwing
Subspecies 4 described

Genus Pyronia

Composed of a few species in Europe and Asia, these butterflies are quite variable in their coloration and patterning.

PYRONIA CECILIA
Southern Gatekeeper

Size	Zone	Status
1¹/₈ in • 30mm	3•4•5	Not protected

Habitat & Ecology grasslands; edges of woods; open clearings in woods and forests
Other Characteristics BELOW similar; FEMALE lacks forewing scent patch
Subspecies more than 8 described

Genus Rhaphicera

This is a small genus of Asian browns with unusual patterns and coloration.

RHAPHICERA MOOREI

Size	Zone	Status
2⁵/₈ in • 65mm	5	Not protected

Habitat & Ecology woods, edges of woods
Other Characteristics ABOVE orange with several brown spots and markings in forewing cell and at apex; hindwing orange with wing base tan and with several ocelli; FEMALE duller
Subspecies none described

Genus Rhinopalpa

This small genus of a single species is characterized by the unusual scalloped and angular wings.

RHINOPALPA POLYNICE EUDOXIA
The Wizard

Size	Zone	Status
3¹/₈ in • 80mm	5•6	Not protected

Habitat & Ecology dense woods at moderate elevations
Other Characteristics BELOW blue white lines on both wings; FEMALE yellowy
Subspecies 10 described

Genus Salamis

Members of this genus are closely related to *Hypolimnas*, and sometimes exhibit several color forms.

♀

SALAMIS ANACARDII
Clouded Mother of Pearl

Size	Zone	Status
3³/₈ in • 85mm	4	Not protected

Habitat & Ecology forests and glades; larval food plant: *Asystasia coromandeliana* (Chinese violet)
Other Characteristics BELOW overscaled with lavender; dull gold ocelli on both wings; MALE with fewer dark markings on wings
Subspecies none described

Genus Sallya

This genus is related to the New World group *Eunica*, and is sometimes considered with it.

▲
♀

SALLYA AMULIA ROSA
Lilac Tree-nymph, Lilac Nymph

Size	Zone	Status
2 in • 50mm	4	Not protected

Habitat & Ecology uncommon species in woods and along streams; attracted to banana bait
Other Characteristics ABOVE lilac with similar spot band on hindwing; MALE above, coloration more vivid
Subspecies 3 described

Genus Siderone

Originally placed in *Anaea*, these butterflies are quite distinctive with the flash patterns above contrasted with the leaf patterns below.

SIDERONE NEMESIS

Size	Zone	Status
2⁷/₈ in • 70mm	2	Not protected

Habitat & Ecology generally humid or mesic tropical rain forest
Other Characteristics BELOW cryptic dark brown; violet lines; FEMALE yellow forewing bands
Subspecies 2 described

Genus Siproeta

This is a small genus composed of 2 species that are widely distributed from Florida to South America.

SIPROETA STELENES BIPLAGIATA

Size	Zone	Status
3 in • 75mm	1•2	Not protected

Habitat & Ecology open woodlands; mesic tropical forests to rain forests; larval food plants: *Ruellia* (petunia), *Justicia* (jacobinia)
Other Characteristics BELOW similar; FEMALE similar
Subspecies 2 described

Genus Speyeria
These are the North American larger fritillaries that are closely related to, and originally incorporated into, *Argynnis*.

SPEYERIA DIANA
Diana

Size	Zone	Status
3⁷/₈ in • 95mm	1	P (in some states)

Habitat & Ecology rich moist woodlands; (southern Appalachian Mountains, Ozarks, USA); mimics Pipevine Swallowtails; larval food plant: *Viola* (violet)
Other Characteristics BELOW similar; silver bar near end cell; FEMALE dimorphic; black with blue lunules in the forewing cell
Subspecies none described

SPEYERIA NOKOMIS
Nokomis Fritillary,
Western Seep Fritillary

Size	Zone	Status
3 in • 75mm	1•2	P (limited)

Habitat & Ecology marshy woodlands and moist meadows; larval food plant: *Viola* (violet)
Other Characteristics ABOVE red orange, darker markings; FEMALE dimorphic; brown with yellow or pale blue submarginal and discal markings
Subspecies 4 described

Genus Stibochiona

These are odd, small nymphalid butterflies that are associated with Asian and Indonesian rain forests.

STIBOCHIONA CORESIA

Size	Zone	Status
2⁷/₈ in • 70mm	5	Not protected

Habitat & Ecology rainforest clearings and paths
Other Characteristics BELOW similar; FEMALE duller
Subspecies 4 described

Genus Stichophthalama

Closely related to *Morpho*, these large butterflies are endemic to Asia.

STICHOPHTHALAMA CAMADEVA
Northern Jungle Queen

Size	Zone	Status
6 in • 150mm	5	Not protected

Habitat & Ecology mesic woods and rain forests; larval food plant: *Poaceae* (bamboo)
Other Characteristics BELOW similar; FEMALE larger
Subspecies 3 described

Genus Taenaris

This is a genus of large brown and white butterflies with large hindwing eyespots. These butterflies are endemic to the Australian region.

TAENARIS DOMITILLA

Size	Zone	Status
4 in • 100mm	6	Not protected

Habitat & Ecology forests and woodlands; larval food plant: *Poaceae* (bamboo)
Other Characteristics ABOVE tan; lighter on posterior half of hindwing; FEMALE paler
Subspecies 2 described

TAENARIS PHORCAS

Size	Zone	Status
4¹/₂ in • 115mm	6	Not protected

Habitat & Ecology infrequently encountered on paths, clearings in rain forests, and forest edges
Other Characteristics BELOW duller; enlarged ocelli; FEMALE more white and smaller hindwing ocelli; lacks black ring around orange patch
Subspecies 2 described

Genus Tanaecia

This genus is composed of more than 50 species, and is endemic to Asia and Indonesia. These butterflies are swift fliers and are related to the genus *Euthalia*.

TANAECIA CLATHRATA

Size	Zone	Status
2⁵/₈ in • 65mm	5	Not protected

Habitat & Ecology woods and edges of woods
Other Characteristics BELOW brown ground color; FEMALE brown with white forewing chevrons; paler brown chevrons on distal hindwing
Subspecies 2 described

Genus Taygetis

The 30 species of large brown butterflies in this genus have distinctive hindwing margins and are associated with mesic tropical and humid rain forests.

TAYGETIS CHRYSOGONE

Size	Zone	Status
4 in • 100mm	2	Not protected

Habitat & Ecology rain forest and tropical forest edges
Other Characteristics BELOW reddish brown ground color; ocelli on both wings; FEMALE similar
Subspecies none described

Genus Tellervo

These black and white butterflies from Australia are closely related to the neotropical *Ithomiines*.

TELLERVO ZOILUS
Hamadryad

Size	Zone	Status
1³/₄ in • 40mm	6	Not protected

Habitat & Ecology primary and mature secondary rain forests to 5,000 ft. (1,500m); larval food plants: *Apocynaceae* (dogbane family)
Other Characteristics BELOW bluish white streaks in forewing cell; FEMALE similar
Subspecies several described

Genus Thaumantis

This is a genus of purple and dark brown butterflies that are closely related to the *Amathusiinae*.

THAUMANTIS DIORES
Jungle Glory

Size	Zone	Status
2 in • 50mm	5	Not protected

Habitat & Ecology rain forest paths; flies at dusk
Other Characteristics BELOW reddish brown; overscaled with white on hindwing; FEMALE duller
Subspecies 2 described

Genus Tirumala

Originally incorporated into the genus *Danaus*, this genus has been subdivided on the basis of structural coloration.

TIRUMALA FORMOSA
Forest Monarch, Beautiful Monarch

Size	Zone	Status
3⁵/₈ in • 90mm	4	Not protected

Habitat & Ecology flies in tropical forests; involved in mimicry complexes and is a model for *Papilio rex*
Other Characteristics BELOW paler; FEMALE duller; lacks hindwing androconial patches
Subspecies 4 markedly different ones described

TIRUMALA LIMNIACE VANERCHENI

Size	**Zone**	**Status**
4¹/₈ in • 105mm	5	Not protected

Habitat & Ecology woodlands; plains; montane areas to 6000 ft (2000m); larval food plant: *Asclepiadaceae* (hoya family)
Other Characteristics BELOW duller; outer margins forewing and hindwing lighter; FEMALE similar
Subspecies 11 described

Genus Tisiphone

This is an Australian genus of large browns that have spectacular markings.

♀

▲

▲

TISIPHONE ABEONA
Sword Grass Brown

Size	**Zone**	**Status**
2⁷/₈ in • 70mm	6	Not protected

Habitat & Ecology rocky meadow areas
Other Characteristics ABOVE dark brown with a narrow yellow orange band enclosing a single large ocellus; MALE similar
Subspecies 7 described

TISIPHONE HELENA
Helene Brown

Size	**Zone**	**Status**
2⁷/₈ in • 70mm	6	Not protected

Habitat & Ecology open woodlands
Other Characteristics ABOVE brown with dark brown forewing apices and a prominent yellow band; FEMALE similar
Subspecies none described

Genus Tithorea
This South American genus is composed of 15 species, and is associated with the *Ithomiinae*.

TITHOREA PINTHIAS

Size	Zone	Status
3⅝ in • 90mm	2	Not protected

Habitat & Ecology humid deciduous tropical rain forests up to 1,500 ft (500m); avid flower visitor
Other Characteristics BELOW yellow postmedian spots near hindwing apex; MALE forewing apices are more sharply pointed
Subspecies 3 described

♀

Genus Vanessa
This is a small group of Admirals that are widely distributed throughout the world. These butterflies are migratory and are strong fliers.

VANESSA ATALANTA
Red Admiral

Size	Zone	Status
2⅝ in • 65mm	1•2•3•5•6	Not protected

Habitat & Ecology open woodlands; edges of woods; open clearings; meadows; very migratory
Other Characteristics BELOW forewing mottled and overscaled with white; FEMALE similar
Subspecies several of doubtful validity described

Genus Vindula

Commonly called the Cruisers, these butterflies have a powerful flight and are often observed at mud puddles or sap flows.

VINDULA ARSINOE ANDEA
Cruiser

Size
4 in • 100mm

Zone
5•6

Status
Not protected

Habitat & Ecology on clearings and paths in rain forests and secondary vegetation
Other Characteristics BELOW similar; FEMALE ground color tan; markings heavier; extra eyespots
Subspecies nearly 20 described

Genus Xanthotaenia

Related to *Stichophthalama*, this genus is composed of a single species from Myanmar to Malaysia and flies at dawn and/or dusk.

XANTHOTAENIA BUSIRIS

Size
3 in • 75mm

Zone
5

Status
Not protected

Habitat & Ecology low flier on trails in open forests up to 3,500 ft (1,200m)
Other Characteristics ABOVE red brown forewing darker at apex, and a white spot with the transverse yellow band; FEMALE similar
Subspecies 4 described

Genus Yoma

This genus is composed of 2 species that are widely distributed throughout Asia and Australia.

YOMA SABINA VASILIA
Australian Lurcher

Size
3¹/₈ in • 80mm

Zone
5•6

Status
Not protected

Habitat & Ecology open woods, forests, and secondary vegetation up to 4,500 ft (1,500m); larval food plant: *Acanthaceae* (acanthia)
Other Characteristics BELOW darker on outer margins; FEMALE paler and larger
Subspecies 5 described

Genus Ypthima

These are small browns or satyrids that are widely distributed in areas of Africa, Asia, and Australia. They have a distinctive mottled appearance below with eyespots on both wings.

YPTHIMA CONJUNCTA

Size | **Zone** | **Status**
$2^3/_8$ in • 60mm | 5 | Not protected

Habitat & Ecology open fields; grassy areas
Other Characteristics ABOVE uniform dark brown; FEMALE larger; above, duller in coloration; dark brown with yellow ringed eyespots at apex or forewing and two at the tornus of hindwing
Subspecies 2 described

Genus Zethera

These are large butterflies that appear to be mimetic of other species in the Asian and Australian regions. Species are sometimes strongly sexually dimorphic.

ZETHERA INCERTA

Size | **Zone** | **Status**
$5^1/_8$ in • 130mm | 6 | Not protected

Habitat & Ecology rain forest; danaid mimic
Other Characteristics ABOVE duskier tan; FEMALE larger
Subspecies none described

ZETHERA MUSA

Size | **Zone** | **Status**
$3^5/_8$ in • 90mm | 5 | Not protected

Habitat & Ecology woodland; rain forest paths and clearings
Other Characteristics ABOVE yellow patch on hindwing; FEMALE dimorphic, can be uniform brown above or brown with similar markings to *Z. incerta*
Subspecies none described

Genus Zeuxidia

Closely related to *Morpho* and *Stichophthalama*, these Asian butterflies display strong sexual dimorphism.

ZEUXIDIA AMETHYSTUS
Saturn

Size	Zone	Status
4¹/₈ in • 105mm	5	Not protected

Habitat & Ecology dense forests near streams up to 4,500 ft (1,360m); drawn to fermenting fruit
Other Characteristics BELOW light brown overscaled with lavender at forewing apex; FEMALE dimorphic; above, brown; yellow forewing band; yellow lunules on hindwing
Subspecies 4 described

ZEUXIDIA AURELIUS

Size	Zone	Status
6⁵/₈ in • 165mm	5	Not protected

Habitat & Ecology dense forests near water
Other Characteristics BELOW dark brown; FEMALE warm brown with numerous cream colored markings and spots on both wings; dimorphic
Subspecies 3 described

CHAPTER 4

Lycaenidae

These are the Gossamer Wings, a family of remarkably diverse, small, yet colorful butterflies. The hairstreaks (Theclinae) are named for the tails on the hindwing margin which may be elongate or absent and often have a "Thecla spot" and or male androconia on the upper forewing. These butterflies generally have rather erratic flight patterns and often perch on the upper sunny surface of leaves. There they will rotate the tails on the hindwing, that appear similar to the antennae on the head. This is similar to a false head, and is a means of avoiding predators inasmuch as these butterflies can survive without a portion of the wing.

The Blues (Polyommatinae) are so named because of the general blue coloration of the wings on the upper surface with remarkably similar pattern elements below.

Genus Agriades

These are small montane butterflies. They are frequently encountered in tundra and in bogs.

AGRIADES GLANDON
Glandon Blue, Primrose Blue

Size	Zone	Status
1¹/₈ in • 30mm	1•3	Not protected

Habitat & Ecology variable, high-elevation sub-arctic bogs; tundra; grassy areas; larval food plant: *Androsace* (rock jasmine)
Other Characteristics ABOVE pale shining greenish blue with darker gray brown marginal borders, especially on forewing apex; FEMALE above, suffused with brown
Subspecies 10 described

Genus Agrodiaetus
This genus is composed of about 20 species from Europe and the Middle East.

AGRODIAETUS AMANDA
Amanda's Blue

Size	Zone	Status
1³/₄ in • 40mm	3•5	Not protected

Habitat & Ecology bogs and moors
Other Characteristics BELOW gray with blue at wing base; black end cell marks and postmedial spot band on forewing; hindwing has two black marks near costa and postmedial spot band FEMALE above, brownish
Subspecies 12 described

AGRODIAETUS THERSITES
Chapman's Blue

Size	Zone	Status
1¹/₈ in • 30mm	3	Not protected

Habitat & Ecology montane meadows
Other Characteristics ABOVE almost uniform purplish blue; FEMALE above primarily brown with purple suffusion at base and orange spots along the submargin of wings
Subspecies more than 10 described

Genus Albulina

This genus of blue butterflies is composed of 6 Eurasian species.

ALBULINA ORBITULUS
Alpine Argus

Size	Zone	Status
1¹/₈ in • 30mm	3	Not protected

Habitat & Ecology high alpine meadows and grassy slopes; larval food plants: *A. alpinus* (alpine calamint), *A. frigidus* (American milkvetch)
Other Characteristics BELOW ground color mid gray; faint black end cell forewing marking and postmedian spot band; hindwing with large cream discal spot and a similar postmedian spot band; FEMALE above, uniform brown
Subspecies none described

Genus Allotinus

This Southeast Asian genus is subdivided into two groups based on the color of the wings' upper surface. Both groups are quite distinctive.

♀

ALLOTINUS APRIES

Size	Zone	Status
1³/₄ in • 40mm	5	Not protected

Habitat & Ecology deep forest
Other Characteristics ABOVE dark brown especially on the forewing costa; MALE similar
Subspecies 3 described

Genus Aloeides

This is a genus of more than 50 species of coppery colored butterflies, most of which occur in South Africa.

ALOEIDES MOLOMO
Molomo Copper

Size	Zone	Status
1¹/₈ in • 30mm	4	Not protected

Habitat & Ecology rocky hillsides; savannahs
Other Characteristics BELOW ground color similar with darker submarginal spot band and fringes; overscaled with white on margin; FEMALE above, similar, with pronounced orange spot band
Subspecies 6 described

Genus Ancema

This genus is composed of more than 4 Southeast Asian species. They are characterized by their rather shiny, silvery undersurface.

ANCEMA BLANKA

Size	Zone	Status
1³/₄ in • 40mm	5	Not protected

Habitat & Ecology infrequently encountered at low and moderate montane areas; males are seen on open, exposed montane slopes
Other Characteristics BELOW silver with hindwing postmedian band and a small thecla spot above first tail; FEMALE paler; above, less iridescent; dark margins more complete
Subspecies 3 described

Genus Anthene

This is a large genus of African and Asian blues commonly called Hairtails.

ANTHENE DEFINITA
Common Hairtail

Size	Zone	Status
1¹/₈ in • 30mm	4	Not protected

Habitat & Ecology dry scrub; gardens; avid flower visitor
Other Characteristics BELOW gray at base; light gray on margins; forewing with rectangular marking end cell and a submarginal spot band; hindwing with two red spots near costa; MALE above, very dark purplish blue
Subspecies 3 described

Genus Antipodilycaena

Endemic to New Zealand, this genus was once considered part of the omnibus genus *Lycaena*.

ANTIPODILYCAENA BOLDENARUM

Size	Zone	Status
1¹/₈ in • 30mm	6	Not protected

Habitat & Ecology submontane meadows
Other Characteristics BELOW dull orange on forewing; gray brown apex; FEMALE dimorphic; above, ground color; dull gold; with iridescent turquoise apical spotband
Subspecies 3 described

Genus Apharitis

Composed of 8 species widely distributed throughout North Africa into southwestern Asia, these butterflies are quite distinct.

APHARITIS ACAMAS TRANSCASPICA
Arab Leopard, Leopard Butterfly

Size	Zone	Status
1³/₈ in • 35mm	3•5	Needs protection

Habitat & Ecology generally desert oases
Other Characteristics BELOW ground color pale yellow; brown markings replaced with iridescent gold, edged in red; FEMALE similar
Subspecies 10 described

Genus Arawacus

This genus of largely Central American hairstreaks has striking lines below.

ARAWACUS JADA
Jade-blue Hairstreak

Size	Zone	Status
1 in • 25mm	2	Not protected

Habitat & Ecology mesic scrub forest
Other Characteristics BELOW creamy yellow to gray; many light bands and stripes; FEMALE similar
Subspecies none described

Genus Arcas

This is a neotropical genus of brightly colored hairstreaks on both surfaces with prominent tails.

ARCAS CYPRIA

Size	Zone	Status
1⁷/₈ in • 45mm	2	Not protected

Habitat & Ecology moist, lowland, tropical rain forests; usually observed on the upper surface of leaves rotating the hindwings in typical hairstreak fashion
Other Characteristics ABOVE brilliant shining iridescent blue; FEMALE above, blue less intense
Subspecies none described

Genus Argyraspodes

Originally associated with *Phasis*, this African genus is composed of a single species, but it is quite distinct.

▲

ARGYRASPODES ARGYRASPIS
Warrior Copper

Size	Zone	Status
1³/₄ in • 40mm	4	Not protected

Habitat & Ecology avid flower visitor in rocky areas; noted for its strong, swift flight
Other Characteristics ABOVE brilliant coppery orange, with narrow black borders; slightly expanded near forewing angle; FEMALE larger
Subspecies none described

Genus Arhopala

Categorized into more than 100 species, these large butterflies are prominently colored above with more cryptic coloration below. Most of the species are distinguished by differences in the coloration of the wings above and below. The critical ndersurface markings on both wings include forewing cell spots, cell end spots, and postmedian, submarginal, and marginal spot bands.

ARHOPALA ARAXES

Size	Zone	Status
2³/₈ in • 60mm	5•6	Not protected

Habitat & Ecology lowland rain forest
Other Characteristics ABOVE brilliant blue with purplish sheen near margins and androconial patch along inner margin; FEMALE pale blue with broader black margins
Subspecies 7 described

ARHOPALA ARGENTEA

Size	Zone	Status
2 in • 50mm	5•6	Not protected

Habitat & Ecology lowland rain forest
Other Characteristics BELOW brown with white scalloped marks on forewing and outer margin of hindwing and more typical of *Arhopala* near base; FEMALE black ground color; cerulean blue scales at wing bases
Subspecies none described

ARHOPALA AUREA

Size	Zone	Status
1⁷/₈ in • 45mm	5	Not protected

Habitat & Ecology along paths and trails in lowland rain forest; both sexes secretive; the males are quick fliers and are generally active in the canopy at dusk
Other Characteristics BELOW dark brown with darker brown markings; FEMALE dimorphic; above, blue with broad black margins
Subspecies 5 described

ARHOPALA CLEANDER

Size	Zone	Status
2 in • 50mm	5•6	Not protected

Habitat & Ecology variable, from dense lowland to montane forests up to 5,000 ft (1,600m)
Other Characteristics BELOW brown; lighter on distal margins of wings with darker brown markings; postdiscal bands on both wings broad and contiguous; FEMALE above, blue with broad black margins
Subspecies 5 described

ARHOPALA MICALE

Size	Zone	Status
2 in • 50mm	6	Not protected

Habitat & Ecology lowland rain forest
Other Characteristics ABOVE metallic blue; with thin, black margins on the wings and a black tail; FEMALE above, ground color primarily black with central area paler blue and duller
Subspecies 16 described

Genus Aricia

This is a genus of blue Eurasian butterflies.

ARHOPALA W. WIDLEYANA

Size	Zone	Status
1 1/8 in • 30mm	5	Not protected

Habitat & Ecology dense lowland forests
Other Characteristics BELOW brown; faint darker brown markings outlined in pale blue on forewing and postmedial band straight along hindwing margins; bluish green patch at hindwing tornus; FEMALE above; paler blue with broader black wing margins
Subspecies none described

Genus Atlides

This genus has 10–12 species of exceedingly gaudy coloration. The red abdomen and markings on the wing bases advertise their toxicity.

ARICIA ARTAXERXES ALLOUS
Mountain Argus, Scotch Argus

Size	Zone	Status
1 1/8 in • 30mm	3	Not protected

Habitat & Ecology generally heaths and montane meadows up to 3,000 ft (1,000m); frequent flower visitor
Other Characteristics ABOVE brown with orange submarginal lunules on both wings; FEMALE duller in coloration
Subspecies 2 described

ATLIDES HALESUS
Great Blue, Great Purple Hairstreak

Size	Zone	Status
1 3/4 in • 40mm	1	Not protected

Habitat & Ecology open woods; forest clearings; larval food plant: *Loranthaceae* (mistletoe family)
Other Characteristics BELOW ground color matte black with three basal carmine red spots and white spots etched in black; iridescent blue markings near hindwing tails; FEMALE duller
Subspecies 3 described

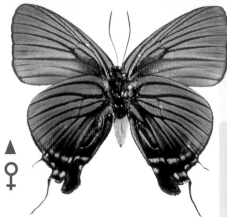

ATLIDES POLYBE

Size	Zone	Status
2¹/₈ in • 55mm	2	Not protected

Habitat & Ecology open woods; clearings; paths; associated with *Loranthaceae* (mistletoe family)
Other Characteristics ABOVE dull bluish gray with broad black margins; MALE similar below; above, much brighter blue; prominent thecla spot and androconial patches on forewing
Subspecies none described

Genus Axiocerces

This African genus is composed of 10 species that occur in tropical and subtropical habitats.

AXIOCERCES BAMBANA
Scarlet Butterfly

Size	Zone	Status
1³/₈ in • 35mm	4	Not protected

Habitat & Ecology scrub forests and often observed hilltopping; avid flower visitors
Other Characteristics BELOW forewing brighter especially at wing base; two gold blue spots in cell, one at end cell, and another posterior to end cell spot with similarly colored postmedian spot band; FEMALE orange ground color; brown subapical and postmedian spot bands
Subspecies none described

Genus Bindahara

This genus has marked sexual dimorphism characterized by extremely long tails. It is restricted to Asia.

BINDAHARA PHOCIDES
Plane Butterfly, Australian Plane

Size	Zone	Status
1¹/₈ in • 30mm	5	Not protected

Habitat & Ecology rain forest
Other Characteristics BELOW forewing gray brown, lighter along inner margin with darker median and postmedian bands; FEMALE above, dark red brown with white lunules on hindwing
Subspecies 8 described

Genus Brephidium

These are exceedingly small butterflies and are commonly called the Pigmy Blues. The genus has representatives in both North and South America and is also found in South Africa.

BREPHIDIUM EXILIS
Western Pygmy Blue, Pygmy Blue

Size	Zone	Status
$3/4$ in • 20mm	1•2	Not protected

Habitat & Ecology mesic to xeric habitats; always near salt flats; larval food plants: *Salvicornia* (saltbush), *Cirsium* (bullthistle)
Other Characteristics ABOVE coppery brown with darker borders; hindwing eyespots not, or only, faintly evident; FEMALE similar
Subspecies 3 described

Genus Caleta

Endemic to Asia and Australia, members of this genus are remarkably similar on both wing surfaces, and interesting studies in black and white.

CALETA MINDARUS

Size	Zone	Status
$1^3/8$ in • 35mm	6	Not protected

Habitat & Ecology usually in forests along streams and in marginal secondary vegetation up to 2,500 ft (800m); infrequently encountered
Other Characteristics ABOVE similar; has a solid black border on lateral margin of the hindwing; FEMALE similar
Subspecies 2 described

Genus Callophrys

This is a genus of small butterflies that are characterized by the green coloration of the undersurface and other structural differences.

CALLOPHRYS AVIS
Chapman's Green Hairstreak

Size	Zone	Status
$1^5/16$ in • 33mm	3	Not protected

Habitat & Ecology usually wood scrub with the larval food plant: *Arbutus* (madrone)
Other Characteristics BELOW ground color emerald with white postdiscal band; MALE above, darker with forewing apices more acute
Subspecies 2 described

CALLOPHRYS RUBI
Green Hairstreak

Size	Zone	Status
1¹⁄₈ in • 30mm	3	Not protected

Habitat & Ecology rough ground in meadows; mountain slopes with brush, heather; larval food plants: *Ulex* (gorse) and *Vaccinium* (blueberry)
Other Characteristics ABOVE unmarked reddish brown; MALE wing apices more acute
Subspecies 8 described

CALLOPHRYS SHERIDANII
White-lined Green Hairstreak,
Sheridan's Hairstreak

Size	Zone	Status
³⁄₄ in • 20mm	1	Not protected

Habitat & Ecology desert to mountain slopes; larval food plant: *Eriogonum* (buckwheat)
Other Characteristics ABOVE; unmarked dark gray brown; MALE forewing apices more acute
Subspecies 4 described

Genus Calycopis

This is a rather large genus of similar-appearing species that are endemic to North and South America.

CALYCOPIS ATRIUS

Size	Zone	Status
1 in • 25mm	2	Not protected

Habitat & Ecology very common forest species
Other Characteristics ABOVE forewings blackish brown; hindwings, silvery blue; FEMALE forewing brown; considerably lighter than male with a small patch of blue along the inner margin; hindwing as in male but duller blue with broader dark margins
Subspecies none described

CALYCOPIS CECROPS
Red-banded Hairstreak

Size	**Zone**	**Status**
1 in • 25mm	1•2	Not protected

Habitat & Ecology variable, from open woods to plains and grassy areas; avid flower visitor
Other Characteristics ABOVE forewings black with extensive cerulean blue on hindwings; FEMALE blue areas reduced
Subspecies none described

Genus Capys

Composed of 4 species that are very brightly colored, this genus is found primarily in South Africa.

Genus Catochrysops

The tailed blues of this genus are widely distributed in the Asian and Australian regions.

CAPYS ALPHEUS
Protea Scarlet

Size	**Zone**	**Status**
1⁷/₈ in • 45mm	4	Not protected

Habitat & Ecology hillsides and rocky terrain; males fly around food plant in late afternoon, and females never venture far from it; larval food plant *Protea* (sugarbush)
Other Characteristics ABOVE brilliant burnt orange with dark brown or black borders on both wings; FEMALE slightly larger; with rounder forewing apices
Subspecies 2 described

CATOCHRYSOPS PANORMUS
Silver Forget-me-not

Size	**Zone**	**Status**
1³/₈ in • 35mm	5•6	Not protected

Habitat & Ecology variable, including wood scrub
Other Characteristics BELOW lighter near tornal margins on wings, with darker gray margins; MALE above, silvery blue and pale in appearance
Subspecies 15 described

Genus Celastrina

This genus comprises more than 40 species that often show seasonal variation.

CELASTRINA ARGIOLUS
Spring Azure, Holly Blue

Size	Zone	Status
1³/₈ in • 35mm	1•2•3	Not protected

Habitat & Ecology open woodlands; wood scrub
Other Characteristics BELOW ground color pearl gray, with darker gray markings; powder blue at base of wings; FEMALE has broader black margins along forewing costa and margins
Subspecies 20 described

Genus Cheritra

The so-called Imperial butterflies in this genus of Asian species have long tails.

CHERITRA FREJA
Common Imperial

Size	Zone	Status
1⁷/₈ in • 45mm	5	Not protected

Habitat & Ecology rain forest paths and trails
Other Characteristics BELOW white overscaled with dull orange on forewing and hindwing apex; FEMALE above, ground color dark brown
Subspecies 9 described

Genus Chilades

This genus of blues is widely distributed in Africa, Asia, and Australia.

CHILADES CLEOTAS

Size	Zone	Status
1³/₄ in • 40mm	6	Not protected

Habitat & Ecology forest edges; secondary and disturbed habitats; populations localized and irregular; larvae attended by ants; larval food plant: *Cycas revoluta* (king sago palm)
Other Characteristics BELOW white; black discal and postdiscal spots on hindwing; FEMALE blue is replaced by warm brown
Subspecies 7 described

Lycaenidae

Genus Chlorostrymon

Divided into 4 species from North and Central America, these butterflies are easily distinguished by the brilliant green on the underwing surfaces.

CHLOROSTRYMON SIMAETHIS
Silver-banded Hairstreak,
St. Christopher's Hairstreak

Size	Zone	Status
³/₄ in • 20mm	1•2	Not protected

Habitat & Ecology open scrub and open areas; larval food plant: *Cardiospermum halicacabum* (love-in-a-puff)
Other Characteristics ABOVE blue to purplish blue shading to black at forewing apex and along lateral margin; FEMALE above, blue purple areas reduced
Subspecies 2 described

Genus Cupido

Composed of 10 species, this genus is restricted to temperate Europe and Asia.

CUPIDO OSIRIS
Osiris Blue

Size	Zone	Status
1¹/₈ in • 30mm	3	Protected

Habitat & Ecology montane meadows between 1,500 and 6,000 ft (500 and 2,000m); larval food plants: *Onobrychis* (sainfoin)
Other Characteristics ABOVE unmarked purplish blue; FEMALE above, ground color unmarked dark brown
Subspecies none described

Genus Curetis

This genus is noted for the striking orange color on the male wings above.

CURETIS FELDERI
Sunbeam Butterfly

Size	Zone	Status
1¹/₈ in • 30mm	5	Not protected

Habitat & Ecology open forests
Other Characteristics BELOW ground color white; very lightly marked; FEMALE above, dull brown with central bands of pale grayish orange on wings; similar below
Subspecies none described

Genus Cyaniris

This is a widespread genus of blues in Europe that occupy a variety of habitats.

CYANIRIS SEMIARGUS
Mazarine Blue

Size	Zone	Status
1³/₈ in • 35mm	3	Not protected

Habitat & Ecology variable; meadow and montane slopes; larval food plants: include *Melilotus officinalis* (clover)
Other Characteristics ABOVE violet blue; FEMALE above, unmarked brown; below, similar
Subspecies 9 described

Genus Cyanophrys

This is a group of neotropical hairstreaks that are characterized by the green ground color on the wings below.

CYANOPHRYS MISERABILIS
Sad Green Hairstreak,
Miserabilis Hairstreak

Size	Zone	Status
1 in • 25mm	1	Not protected

Habitat & Ecology forest edges; wood scrub; disturbed habitats; populations localized
Other Characteristics ABOVE unmarked grayish blue; FEMALE duller in coloration
Subspecies 2 described

Genus Dacalana

Characterized by the two hindwing tails, these butterflies have been divided into two groups based on structural and habitat differences.

DACALANA VIDURA AZYADA
Double Tufted Royal

Size	Zone	Status
1³/₈ in • 35mm	5	Not protected

Habitat & Ecology variable; one group on rain forest trails up to moderate elevations; the other group infrequently in montane clearings
Other Characteristics ABOVE bright blue with black forewing apex; FEMALE similar
Subspecies 2 described

Lycaenidae

Genus Danis
With more than 12 species, these butterflies are recognized by the characteristic wing patterns on the undersurface.

DANIS DANIS
Large Green-banded Blue

Size	Zone	Status
1³/₄ in • 40mm	6	Not protected

Habitat & Ecology variable, open primary or mature secondary rain forest
Other Characteristics ABOVE pale uniform blue with the white band below evident; FEMALE above, blackish brown with reduced white bands on both wings
Subspecies 15 described

Genus Deudorix
There are between 27 and 80 species recognized in this genus. These butterflies are widely distributed in the African, Asian, and Australian regions.

DEUDORIX DIOCLES
Orange-barred Playboy

Size	Zone	Status
1¹/₈ in • 30mm	4	Not protected

Habitat & Ecology variable; paths and trails in open woodlands; coastal scrub; and grasslands
Other Characteristics BELOW gray brown ground color; gray spots and postmedian band; FEMALE dimorphic; brown, white, and blue discal patches
Subspecies none described

DEUDORIX EPIRUS EOS
Blue Cornelian

Size	Zone	Status
1⁷/₈ in • 45mm	6	Not protected

Habitat & Ecology variable but generally open woodlands or mature secondary forest
Other Characteristics BELOW ground color black; white overscaled with black at wing bases; broad white postdiscal band on both wings; FEMALE above, blackish with blue at base
Subspecies 5 described

▲
♀

DEUDORIX ERYX

Size
2³/₈ in • 60mm

Zone
5

Status
Not protected

Habitat upland and lowland rain forest
Other Characteristics ABOVE ground color brown;
white patch along margin from the tornus to
middle of hindwing; MALE dimorphic; above, vivid
blue with broad black margins and apices
Subspecies 4 described

Genus Eumedonia

Composed of 4 species, this genus is
widely distributed in Europe and Asia.

Genus Eliotia

This genus of Asian hairstreaks is
closely related to *Myrina*, *Tajuria*, and
the genera in the *Jacoona* group.

▲♀

ELIOTIA JALINDRA

Size
1³/₄ in • 40mm

Zone
5

Status
Not protected

Habitat & Ecology open forest and gardens at
low to moderate elevations
Other Characteristics ABOVE ground color dark
brown overscaled sparsely with white;
MALE dimorphic; above, ground color deep blue
Subspecies 10 described

EUMEDONIA EUMEDON
Geranium Argus

Size
1⁷/₈ in • 45mm

Zone
3

Status
Not protected

Habitat & Ecology lowland and montane
meadows, rocky slopes to 8,000 ft.(2,400m);
avid flower visitor; larval food plant: *Geranium
praetense* (meadow geranium)
Other Characteristics ABOVE uniformly gray
brown; FEMALE above and below, similar to male,
but occasionally with orange lunules on upper
hindwing submargin
Subspecies 15 described

Genus Euphilotes

The 4 species in this group are restricted to western North America and are associated with *Eriogonum* (buckwheat) as larvae.

EUPHILOTES RITA
Rita Blue, Desert Buckwheat Blue

Size	Zone	Status
1 in • 25mm	1	Not protected

Habitat & Ecology mesic sagebrush areas
Other Characteristics ABOVE ground color brilliant blue; rather unmarked; FEMALE above, brown with orange submarginal band on hindwing; similar below
Subspecies 6 described

Genus Evenus

These large hairstreak butterflies are usually brilliant blue above with unique, colorful patterns below.

EVENUS TERESINA

Size	Zone	Status
7/8 in • 20mm	2	Not protected

Habitat & Ecology tropical and subtropical rain forests
Other Characteristics ABOVE ground color vivid blue; FEMALE above, ground color lighter; dark margins and forewing apices; below, broader red markings on hindwing
Subspecies none described

Genus Everes

A few representatives of this genus, comprising 12 species, 1 of which is found in Latin America.

EVERES COMYNTAS
Eastern Tailed Blue

Size	Zone	Status
1 in • 25mm	1•2	Not protected

Habitat & Ecology open country species, often found in parks and gardens; larval food plants: include *Fabaceae* (legume/pea family)
Other Characteristics BELOW ground color white with faint grayish black markings; up to 3 orange lunules on hindwing margin; FEMALE above, ground color uniform blackish brown with a few blue scales at base of wings in spring forms
Subspecies 2 described

Genus Fixenia

These hairstreaks are widely distributed in Europe, Asia, and North America.

FIXENIA PRUNI
Black Hairstreak

Size	Zone	Status
1¹/₈ in • 30mm	3•5	Protected

Habitat & Ecology open woodlands
Other Characteristics ABOVE ground color uniform dark brown with dull orange shading near hindwing border; FEMALE similar
Subspecies 2 described

Genus Glaucopysche

The 10 species in this genus are noted for their bright blue coloration. They are widely distributed in temperate Europe, Asia, and North America.

Genus Flos

This genus of more than 12 species is endemic to Southeast Asia.

FLOS ANNIELLA

Size	Zone	Status
1⁷/₈ in • 45mm	3	Not protected

Habitat & Ecology forest paths near streams
Other Characteristics BELOW ground color reddish brown with darker brown markings outlined in greenish blue; iridescent edge near hindwing tails; FEMALE above, cerulean blue with broad dark margins on forewing and hindwing
Subspecies 3 described

GLAUCOPYSCHE ALEXIS
Green-underside Blue

Size	Zone	Status
1³/₈ in • 35mm	5	Not protected

Habitat & Ecology generally montane open woods or meadows with flowers up to 4,000 ft (1,200m); larval food plant: various *Fabaceae* (legume/pea family)
Other Characteristics ABOVE ground color brilliant shining blue with well-defined black margins; FEMALE above, ground color brown with some blue shading toward base of wings
Subspecies 12 described

GLAUCOPYSCHE LYGDAMUS AFRA
Silvery Blue

Size	Zone	Status
1¹/₈ in • 30mm	1	Protected

Habitat & Ecology variable, and widespread in open woodlands, montane areas, scrub; larval food plants: *Fabaceae* (legume/pea family)
Other Characteristics BELOW silvery gray to gray, with a row of postmedian black spots on forewing outlined in white; FEMALE above, brown, with blue overscaling toward wing bases
Subspecies 12 described

Genus Habrodais

This genus comprises 2 North American species that are restricted to the West Coast.

HABRODAIS GRUNUS
Live-oak Hairstreak, Golden Hairstreak

Size	Zone	Status
1¹/₈ in • 30mm	1	Not protected

Habitat & Ecology open oak woods; adults generally remain in trees near larval food plants: *Fagaceae* (oak family)
Other Characteristics BELOW ground color golden brown with faint indication of median, and submarginal bands, and pale metallic crescents near margin; FEMALE similar
Subspecies 3 described

Genus Harkenclenus

These tailless hairstreaks are widely distributed over much of North America.

HARKENCLENUS TITUS
Coral Hairstreak

Size	Zone	Status
1¹/₈ in • 30mm	1	Not protected

Habitat & Ecology variety of open woods
Other Characteristics ABOVE ground color rich brown with orange marginal band on hindwing; FEMALE similar
Subspecies 4 described

HEMIARGUS THOMASI
Caribbean Eyed Blue, Miami Eyed Blue, Miami Blue, Thomas's Eyed Blue

Size	Zone	Status
1 1/8 in • 30mm	1•2	Protected

Habitat & Ecology variable from moist to mesic open woodlands and along forest edges
Other Characteristics ABOVE ground color bright blue, darker at forewing apex and along lateral margin; hindwing with indication of submarginal spotband; FEMALE above, ground color brown, faintly suffused with blue near wing base, two prominent eyespots near hindwing tornus
Subspecies 5 described

Genus Hemiargus

Widely distributed in the southern United States, the West Indies, and Central and South America, this is a genus of blues commonly called the Eyed Blues.

Genus Heodes

These coppers are restricted to the temperate areas of Europe and Asia.

♀

HEODES TITYRUS
Sooty Copper

Size	Zone	Status
1 1/8 in • 30mm	3	Not protected

Habitat & Ecology flowering meadows and fallow fields where *Rumex* (dock) is found
Other Characteristics BELOW forewing ground color orange, with gray along costal and lateral margins; hindwing grayish white with median, postmedian, and submarginal spots; prominent orange marginal band; MALE dimorphic; above, sooty ground color
Subspecies 7 described

HEODES VIRGAUREAE
Scarce Copper

Size	Zone	Status
1 3/8 in • 35mm	3	Not protected

Habitat & Ecology variable, flowering meadows from lowland to 5,000 ft (1,600m)
Other Characteristics BELOW forewing suffused with yellow orange with black markings in cell, and black postmedian and submarginal spot bands; FEMALE above, ground color copper; heavy black markings on both wings
Subspecies 20 described

Genus Horaga

This is a genus of Asian and Australian hairstreaks that inhabit a variety of secondary forests.

HORAGA ONYX FRUHSTORFERI

Size	Zone	Status
1³/₈ in • 35mm	5•6	Not protected

Habitat & Ecology variety of lowland and upland secondary forests, and exposed hilltops

Other Characteristics BELOW ground color warm reddish brown, darker near forewing apex with broad linear, irregularly shaped white bands outlined in black on both wings; FEMALE more extensive dark markings with iridescent blue reduced on wings

Subspecies 9 described

Genus Hypaurotis

This genus comprises a single species, and is widely distributed in western North America.

HYPAUROTIS CRYSALUS
Colorado Hairstreak

Size	Zone	Status
1³/₄ in • 40mm	1	Not protected

Habitat & Ecology normally scrub oak habitats

Other Characteristics BELOW ground color light to dark gray with prominent white edged in black on postmedian band; FEMALE black bar in end cell extends to forewing tornus

Subspecies 2 described

Genus Hypochlorosis

Once considered part of *Hypochrysops*, these are found in the Australian region.

▲

HYPOCHLOROSIS PAGENSTECHERI SCINTILLANS

Size	Zone	Status
1³/₈ in • 35mm	6	Not protected

Habitat & Ecology paths and rain forest edges

Other Characteristics ABOVE ground color iridescent blue with extensive black at forewing apex; FEMALE above, ground color warm brown; darker on forewing apex with duller iridescent blue scales in forewing cell and areas posterior to cell

Subspecies 2 described

Genus Hypochrysops
This genus of brightly colored butterflies with distinctive undersurface wing color patterns is restricted to the Australian region.

HYPOCHRYSOPS APELLES
Copper Jewel

Size	Zone	Status
1 1/8 in • 30mm	6	Not protected

Habitat & Ecology coastal mangrove swamps; sand dunes; subcoastal savanna woodland
Other Characteristics BELOW ground color on forewing muted orange at base; hindwing with dark red bands; FEMALE ground color is duller
Subspecies 2 described

HYPOCHRYSOPS POLYCLETUS
Rovena Jewel

Size	Zone	Status
1 3/4 in • 40mm	6	Not protected

Habitat & Ecology swampy forests; marginal secondary forests up to 1,600 ft (500m)
Other Characteristics BELOW ground color brown; red bands and spots outlined in black and iridescent green in and anterior to forewing cell; FEMALE markedly dimorphic; above, ground color dark brown; basal blue scales on forewing
Subspecies 12 described

Genus Hypolycaena
This genus is found across the African, Asian, and Australian regions.

HYPOLYCAENA BUXTONI
Buxton's Hairstreak

Size	Zone	Status
1 1/8 in • 30mm	4	Not protected

Habitat & Ecology open woods; savannahs
Other Characteristics ABOVE ground color bluish purple; FEMALE above, ground color blackish brown; with broad white median forewing patch; white postmedian, and thin submarginal spot bands on hindwing
Subspecies 4 described

HYPOLYCAENA DANIS
Black and White Tit

Size	Zone	Status
1 1/8 in • 30mm	6	Not protected

Habitat & Ecology open forest; forest edges; larval food plant: *Dendrobium* (rock lily)
Other Characteristics BELOW margins brownish black and overscaled with yellow on hindwing white patch; marginal blue markings and eyespots enlarged; FEMALE above, forewing apices rounded and with expanded blue on hindwing submargins
Subspecies 5 described

HYPOLYCAENA ERYLUS TEATUS
Common Tit

Size	Zone	Status
1 3/8 in • 35mm	6	Not protected

Habitat & Ecology variable from forests and forest edges to open areas
Other Characteristics BELOW ground color tan gray with darker submarginal bands on both wings; hindwing with darker postmedian band and overscaled with white along submargin; FEMALE above, ground color brown; occasionally with ochreous patch in forewing discal area
Subspecies 6 described

Genus Icaricia

The arctic blues in this species are usually sexually dimorphic.

ICARICIA ACMON
Emerald-studded Blue,
Silver-studded Blue

Size	Zone	Status
1 in • 25mm	1	Not protected

Habitat & Ecology variable, sagebrush flats and washes; sunny forest trails; arid montane slopes; adults fly close to the ground and congregate at mud puddles
Other Characteristics ABOVE bright blue with narrow orange submarginal band on hindwing and black submarginal spots; FEMALE above, ground color brown with forewing orange bars on submargin and tornus
Subspecies 4 described

Genus Incisalia
Subdivided into 12 species, these Elfin butterflies are various shades of brown on the upper wing surface and cryptically colored below.

▲

▲ ♀

INCISALIA NIPHON
Eastern Pine Elfin

Size	Zone	Status
1 1/8 in • 30mm	1	Not protected

Habitat & Ecology usually pine woodlands; larval food plant: *Pinus banksiana* (jack pine)
Other Characteristics ABOVE ground color usually brown, occasionally with russet overtones; FEMALE similar markings but more red overall
Subspecies 2 described

INCISALIA POLIOS
Hoary Elfin

Size	Zone	Status
1 in • 25mm	1	Not protected

Habitat & Ecology forest and woods; larval foodplant: *Arctostaphylos uvaursi* (bearberry)
Other Characteristics ABOVE ground color, dull brown; MALE similar
Subspecies 2 described

Genus Iolana
Comprising Eurasian and North African species, this is a genus of large blues.

IOLANA IOLAS
Iolas Blue

Size	Zone	Status
1 3/4 in • 40mm	3	Not protected

Habitat & Ecology open woodlands, meadows, and rocky areas where *Colutea istria* (bladder senna) occurs
Other Characteristics BELOW silvery gray with black submarginal forewing spot band; hindwing with a black postdiscal spot band; faint gray marginal bands on both wings; FEMALE ground color blue shading to purple near wing margins; similar below
Subspecies 15 described

Genus Iolaus
The hairstreaks in this African genus are quite distinct in wing patterns and coloration. As larvae, these butterflies are associated with *Loranthus* (mistletoe), which affords them some protection from predators.

IOLAUS BOWKERI
Bowker's Tailed Blue

Size	Zone	Status
1³/₄ in • 40mm	4	Not protected

Habitat & Ecology variable, but generally associated with scrub and brush areas along forest edges and margins
Other Characteristics BELOW ground color gray, overscaled with rust on both wing margins; FEMALE above, similar and below with white markings on wings slightly expanded
Subspecies 10 described

IOLAUS SIDUS
Red-line Sapphire Blue

Size	Zone	Status
1³/₈ in • 35mm	4	Not protected

Habitat & Ecology variable, open woodlands; forest clearings and edges; coastal forests; savannah
Other Characteristics BELOW ground color tan gray with dark red linear postmedian band outlined in black near forewing subapex; FEMALE similar but expanded black margins on wings
Subspecies none described

Genus Jacoona
This is a genus composed of 3 Asian species noted for their elongated hindwing tails.

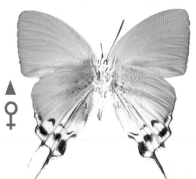

JACOONA AMRITA
Grand Imperial

Size	Zone	Status
2 in • 50mm	5	Not protected

Habitat & Ecology low- and upland open forests
Other Characteristics ABOVE brown with a white patch on hindwing near margin 2 black spots above tails; MALE dimorphic; above, ground color iridescent blue; broad forewing apex and margins; similar markings near tails
Subspecies 2 described

Genus Jalmenus

This genus is endemic to Australia and associated with *Acacia* (mimosa tree) as larvae.

JALMENUS EVAGORAS
Common Imperial Blue

Size	Zone	Status
1³/₈ in • 35mm	6	Not protected

Habitat & Ecology variable; open woodlands; scrub; secondary habitats
Other Characteristics BELOW pale yellow; discal black dashes on basal half; brown and tan marginal area on both wings; FEMALE similar
Subspecies 2 described

Genus Jamides

With more than 50 blues, this genus is distributed throughout Asia and Indonesia, east to Fiji and the Tonga Islands.

▲
♀

JAMIDES ABDUL

Size	Zone	Status
1³/₈ in • 35mm	5	Not protected

Habitat & Ecology open woods and forest paths
Other Characteristics ABOVE ground color dull blue with black along forewing margin; hindwing with black submarginal bands; MALE dimorphic; above, brilliant greenish blue, with broad black forewing margins on wings
Subspecies 4 described

JAMIDES CUNILDA

Size	Zone	Status
1³/₈ in • 35mm	5	Not protected

Habitat & Ecology open woodlands and forest paths; secondary growth
Other Characteristics BELOW ground color gray, brown with linear broken stripes on outer half of forewing, and the entire surface of hindwing; enlarged eyespot with broad orange patch on hindwing; FEMALE above, ground color paler blue; broad black on forewing apex, lateral margin; black submarginal spot band
Subspecies 3 described

Genus Laeosopis

This unusual genus of hairstreaks is distinctively purple on the upper surface.

LAEOSOPIS ROBORIS
Spanish Purple Hairstreak

Size	Zone	Status
1¹/₈ in • 30mm	3	Not protected

Habitat & Ecology open woodland and woodland edges up to 5,000 ft (1,500m) where *Fraxinus* (ash) grows
Other Characteristics ABOVE ground color purple at base and in disc; with broad black margins on wings; FEMALE above, purple areas restricted
Subspecies 9 described

Genus Lampides

This is a strongly migratory genus that is widespread on continents throughout most of the world.

LAMPIDES BOETICUS
Long-tailed Blue

Size	Zone	Status
1¹/₈ in • 30mm	1•3•4•5•6	Not protected

Habitat & Ecology variable, from paths and trails in secondary forests to agricultural crops
Other Characteristics ABOVE ground color pale blue with darker apex and margins on wings; MALE above, ground color bright blue to violet blue; darker narrow border, and with numerous androconia; below, darker
Subspecies 2 described

Genus Leptomyrina

This genus of 9 species has very unusual wing markings.

LEPTOMYRINA GORGIAS
Black Eye

Size	Zone	Status
1³/₈ in • 35mm	4	Not protected

Habitat & Ecology infrequently seen in scrub vegetation, especially in montane rocky outcrops
Other Characteristics BELOW ground color gray brown, white along forewing margin, apex, and lateral margin; FEMALE more rounded forewings
Subspecies none described

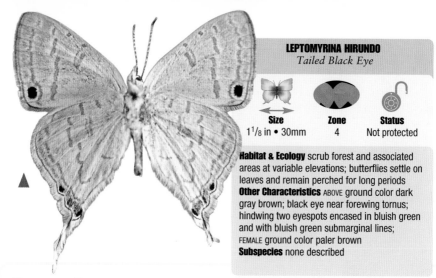

LEPTOMYRINA HIRUNDO
Tailed Black Eye

Size
1 1/8 in • 30mm

Zone
4

Status
Not protected

Habitat & Ecology scrub forest and associated areas at variable elevations; butterflies settle on leaves and remain perched for long periods
Other Characteristics ABOVE ground color dark gray brown; black eye near forewing tornus; hindwing two eyespots encased in bluish green and with bluish green submarginal lines; FEMALE ground color paler brown
Subspecies none described

Genus Leptotes
This is a pantropical genus of 17 species that have a distinctive pattern on the undersurface. These butterflies are strongly migratory.

LEPTOTES MARINA
Marine Blue, Striped Blue

Size
1 in • 25mm

Zone
1•2

Status
Not protected

Habitat & Ecology open areas; mesquite thorn scrub; sparse woodlands or edges of woods
Other Characteristics BELOW ground color white with gray brown bars and bands; blackish gray marginal spot band with two black eyespots capped with greenish blue; FEMALE above, ground color bright blue with broader black margins
Subspecies none described

LEPTOTES PIRITHOUS
Lang's Short-tailed Blue

Size
1 in • 25mm

Zone
3•4•5

Status
Not protected

Habitat & Ecology common in Africa; found in all altitudes; larval food plant: *Plumbago* (leadwort)
Other Characteristics ABOVE almost unmarked blue, shading to white; broad dark margins on forewing; white submarginal spot on hindwing; FEMALE similar
Subspecies 3 described

Genus Liphyra

This is an unusal genus composed of 2 rather large species that are quite different in appearance for members of the Lycaenidae.

♀

LIPHYRA BRASSOLIS
Moth Butterfly

Size	Zone	Status
3⁵/₈ in • 90mm	5•6	Not protected

Habitat & Ecology paths and trails in dense forest
Other Characteristics ABOVE ground color yellow orange; forewing has broad dark brown apex and lateral margin; hindwing has dark brown spots outside cell; MALE dark brown on wing margins; more extensive markings
Subspecies 10 described

Genus Loxura

With their distinctive coloration and elegant long tails, these butterflies are easily recognizable.

LOXURA ATYMNUS
Yam Fly

Size	Zone	Status
1³/₄ in • 40mm	5	Not protected

Habitat & Ecology open, disturbed areas; pest; larval food plants: include *Dioscorea* (yam)
Other Characteristics BELOW ground color with darker mottling at base; similarly colored spotbands; FEMALE above, ground color pale yellow orange, broader dark margins
Subspecies 16 described

Genus Lycaeides

Widely distributed in North America, Europe, and Asia, this genus has orange submarginal spot bands on the uppersurface of the hindwing.

LYCAEIDES MELISSA
Orange-margined Blue

Size	Zone	Status
1¹/₈ in • 30mm	1	Not protected

Habitat & Ecology any open habitat
Other Characteristics ABOVE ground color blue with orange submarginal bands; FEMALE dimorphic; above, brown with blue overscaling on base of wings
Subspecies 5 described

Genus Lycaena

This rather inclusive genus of coppers is present in some form on almost every continent.

LYCAENA FEREDAYI

Size	Zone	Status
1 1/8 in • 30mm	6	Not protected

Habitat & Ecology open meadows and grassy areas with flowers
Other Characteristics BELOW ground color yellowish orange with markings as above and a complete submarginal spot band; hindwing yellow gray with small dark spots at base and in cell, and a complete postmedian spot band; FEMALE lacks violet suffusion; with more distinct markings
Subspecies none described

LYCAENA EPIXANTHE
Cranberry-bog Copper, Bog Copper

Size	Zone	Status
1 in • 25mm	1	Not protected

Habitat & Ecology bogs
Other Characteristics ABOVE ground color brown with a purplish sheen; FEMALE above, more spots
Subspecies 4 described

LYCAENA HELLE
Violet Copper

Size	Zone	Status
1 1/8 in • 30mm	3	Not protected

Habitat & Ecology moist flowery meadows and marshes to about 5,000 ft (1,500m)
Other Characteristics ABOVE ground color glossy purple with an orange hindwing submarginal band; FEMALE above, blackish; yellow bars along outer part of forewing and along hindwing submargin; below, paler
Subspecies 12 described

LYCAENA HETERONEA
Blue Copper

Size	Zone	Status
1³/₈ in • 35mm	1	Not protected

Habitat & Ecology open meadows; sagebrush-area slopes; avid flower visitor
Other Characteristics BELOW ground color silvery white or gray with or without faint black markings; FEMALE above, brown, usually with no indication of any blue at wing base, although a few scales may be present; 2 rows of black medial and postmedial spot bands, especially on the forewing
Subspecies 2 described

LYCAENA HYLLUS
Bronze Copper

Size	Zone	Status
1³/₄ in • 40mm	1	Not protected

Habitat & Ecology meadows, especially moist areas where *Rumex* (dock) occurs
Other Characteristics BELOW forewing orange at base, white at apex with black markings in cell and black postdiscal and submarginal spot bands; blue gray hindwing; FEMALE above, ground color bright orange with black discal spots and a dark margin; dark gray hindwing
Subspecies none described

LYCAENA ORUS
Sorrel Copper

Size	Zone	Status
1¹/₈ in • 30mm	4	Not protected

Habitat & Ecology variable from coastal to montane meadows where *Rumex* (dock) occurs
Other Characteristics BELOW forewing with paler orange at base shading to brown mottled with white at apex; markings similar; hindwing brown, mottled heavily with white at base; FEMALE above, ground color slightly duller
Subspecies none described

♀

LYCAENA PHLAEUS
Small Copper, American Copper

Size	Zone	Status
1¹/₈ in • 30mm	3•4	Not protected

Habitat & Ecology open meadows near woods
Other Characteristics BELOW forewing apex and hindwing margin gray brown; MALE forewing apex more acute
Subspecies 20 described

LYCAENA SALUSTIUS

Size	Zone	Status
1¹/₈ in • 30mm	6	Not protected

Habitat & Ecology warm flowery meadows
Other Characteristics BELOW forewing ground color orange, darker on margins with reduced dark brownish black markings as above; hindwing golden, overlaid with pink on margins with faint indication of reduced markings; FEMALE above, ground color brown, shading to yellow on submarginal areas of wings; below, darker; hindwing brown with pink overlaid on outer half of wing
Subspecies none described

Genus Lysandra

This genus of blues, consisting of more than 20 species, is endemic to the temperate areas of Europe and Asia.

LYSANDRA BELLARGUS
Adonis Blue

Size	Zone	Status
1¹/₈ in • 30mm	3	Not protected

Habitat & Ecology calcareous lowland meadows and montane slopes to 6,000 ft (2,000m)
Other Characteristics BELOW ground color light gray brown with black markings; black spot band on hindwing, capped with orange; FEMALE above, brown, with orange submarginal hindwing spot band
Subspecies 12 described

LYSANDRA CORIDON
Chalk-hill Blue

Size	Zone	Status
1³/₈ in • 35mm	3	Not protected

Habitat & Ecology associated with chalk and limestone and usually found in grassy areas
Other Characteristics BELOW forewing gray with fine black markings and incomplete gray submarginal spot band; hindwing gray brown, darker toward margin with white patches in cell toward apex and along mid-postmedian area; FEMALE above, dark gray; orange margins more extensive
Subspecies 12 described

Genus Maculinea

This genus includes the temperate large blues of Europe and Asia. The caterpillars and pupae are tended by ants and often produce secretion and sounds to attract them.

MACULINEA ARION
Alcon Blue, Large Blue

Size	Zone	Status
1³/₄ in • 40mm	3	Protected

Habitat & Ecology rough grassy areas where *Thymus serpyllum* (thyme) occurs; larva pupates in ant nests
Other Characteristics BELOW ground color gray brown with bluish violet flush on wing bases and faint darker marginal spot band on wings; full complement of fine spots in cell, postmedian, and submarginal spot bands; FEMALE above, ground color gray brown, darker toward margins with bluish purple overscaling at base of wings
Subspecies 25 described

MACULINEA TELEIUS
Scarce Large Blue

Family	Zone	Status
1³/₈ in • 35mm	3	Vulnerable

Habitat & Ecology moist meadows and marshes where *Sanginsorba officinalis* (great burnet) grows
Other Characteristics BELOW ground color pale brown; blue markings finer and fainter; hindwing lacks bluish green basal flush; FEMALE above, duller; more markings on forewing
Subspecies 12 described

Genus Mahathala

This genus is closely related to the genus *Arhopala*, but it differs in the shape of the hindwing costa and tails. It is widely distributed from India eastward to China and Sundaland.

MAHATHALA AMERIA HAINANI

Size	**Zone**	**Status**
1³/₄ in • 40mm	5	Not protected

Habitat & Ecology forest trails and paths; often in mature secondary growth
Other Characteristics BELOW ground color warm brown at base and darker on margins; forewing overscaled with white at apex, at hindwing base, and outer half of wing; FEMALE above, duller; dark blue purple
Subspecies 12 described

Genus Manto

This genus has an erect hair tuft on the hindwing's inner margin and the tails.

MANTO HYPOLEUCA

Size	**Zone**	**Status**
1³/₄ in • 40mm	5	Not protected

Habitat & Ecology forest dwellers at low to moderate elevations; infrequently encountered
Other Characteristics BELOW ground color gold, paler on lower half of wings; lustrous white patch near inner margin of forewing; FEMALE dimorphic; above, ground color rich brown; paler below
Subspecies 6 described

Genus Mantoides

Wing venation distinguishes this genus.

MANTOIDES GAMA

Size	**Zone**	**Status**
1⁷/₈ in • 45mm	5	Not protected

Habitat & Ecology encountered infrequently in forests at low to moderate elevations
Other Characteristics BELOW forewing ground color gold darkening to orange or copper at apex along costal and lateral margins; yellowish white hindwing; FEMALE reduced white area at hindwing tornus
Subspecies 3 described

Genus Megisba

The small blue butterflies of this genus are widely distributed in the Asian and Australian regions.

Size	Zone	Status
1 in • 25mm	5•6	Not protected

Habitat & Ecology paths and trails in open forests and plains; females secretive; larval food plant: *Sapindaceae* (soapberry family)
Other Characteristics BELOW ground color warm brown, darker at forewing apex; forewing with angular white patch below cell and faint white diffuse submarginal spot band; FEMALE similar
Subspecies 12 described

Genus Micandra

This neotropical genus, found mostly in subandean areas of South America, has lustrous blue above.

▲
♀

Genus Mimacraea

These lycaenids have similar markings and are mimetic of the African *Acraea*, some danaids, and nymphalids.

MICANDRA PLATYPTERA

Size	Zone	Status
1⅛ in • 30mm	2	Not protected

Habitat & Ecology open moist montane forests
Other Characteristics ABOVE ground color duller blue, with broad brownish black margins, especially on forewing; MALE dimorphic; with extremely round wings; lacks tail; above, lustrous blue; forewing with darker blue androconial patch in cell near apex
Subspecies none described

MIMACRAEA MARSHALLI DOHERTYI
Marshall's Acraea Mimic

Size	Zone	Status
2⅛ in • 55mm	4	Not protected

Habitat & Ecology open woodlands
Other Characteristics BELOW ground color tan orange with limited black on forewing as above and along both wing margins; FEMALE somewhat larger but forewings rounder; numerous black spots on hindwing
Subspecies 4 described

Genus Ministrymon

Quite small, these butterflies are endemic to the southwestern United States and into Mexico.

♀ ▲

MINISTRYMON CLYTIE
Silver-blue Hairstreak

Size	Zone	Status
⁷/₈ in • 20mm	1•2	Not protected

Habitat & Ecology mesic habitats, such as desert washes near *Prosopis juliflora* (mesquite)
Other Characteristics ABOVE ground color brown with blue on the forewing at base and along inner margin; MALE above, ground color nearly uniform brown with blue at base of wings
Subspecies none described

Genus Mitoura
As larvae this genus feeds on some sort of conifers, cedar, and mistletoe.

▲

♀ ▲

MITOURA GRYNEUS
Cedar Hairstreak, Olive Hairstreak

Size	Zone	Status
1 in • 25mm	1	Not protected

Habitat & Ecology generally open red cedar woodlands; larval food plants: include *Juniperus virginiana* (pencil cedar)
Other Characteristics ABOVE ground color dark brown, sometimes overscaled with orange along the outer wing margins; FEMALE above, with larger orange marks
Subspecies 4 described

MITOURA SPINETORUM
Blue Mistletoe Hairstreak

Size	Zone	Status
1 in • 25mm	1	Not protected

Habitat & Ecology various open woods in mountains or hilly areas
Other Characteristics ABOVE ground color dark steely blue; MALE similar
Subspecies 2 described

Genus Myrina

These African hairstreaks are distinguished by the shape and length of their hindwing tails.

MYRINA SILENUS
Figtree Blue

Size	Zone	Status
1³/₄ in • 40mm	4	Not protected

Habitat & Ecology trails and clearings in woodlands where figs grow
Other Characteristics BELOW ground color pale rust tan, lighter on outer margins with faint indication of postmedian bands on wings; FEMALE above, duller coloration; below, ground color darker
Subspecies 6 described

Genus Neomyrina

Similar to *Myrina*, this genus is characterized by hindwing shape and white coloration on the wing surfaces.

NEOMYRINA NIVEA
White Imperial

Size	Zone	Status
2³/₈ in • 60mm	5	Not protected

Habitat & Ecology forest paths and edges
Other Characteristics BELOW ground color white; darker along margins; with mottled grayish linear bands; FEMALE larger; forewing more rounded at apex
Subspecies 3 described

Genus Neopithecops

These blues are endemic to the Asian and Australian regions.

NEOPITHECOPS ZALMORA

Size	Zone	Status
⁷/₈ in • 20mm	5•6	Not protected

Habitat & Ecology woodland; lowland forest paths; frequently encountered; larval food plant: *Glycosmis pentaphylla* (Rutaceae family)
Other Characteristics ABOVE ground color dark brown with the forewing medial area overscaled with white; FEMALE similar
Subspecies 8 described

Genus Oenomaus

This is a genus of large Central and South American hairstreaks.

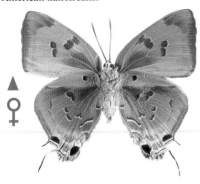

♀ ▲

OENOMAUS RUSTAN

Size	Zone	Status
1⁷⁄₈ in • 45mm	2	Not protected

Habitat & Ecology open tropical forests and forest edges
Other Characteristics ABOVE ground color brown with blue gray overscaling on forewing base and almost entire hindwing; MALE above, ground color brown; central forewings blue above inner margin
Subspecies none described

Genus Ogyris

Categorized into 15 brightly colored species, these butterflies are endemic to the Australian region.

▲

♀

OGYRIS AMARYLLIS HEWITSONI
Amaryllis Azure

Size	Zone	Status
1⁷⁄₈ in • 45mm	6	Not protected

Habitat & Ecology trails and paths in open woodlands; larval food plant: *Phoradendron* (mistletoe)
Other Characteristics ABOVE brilliant blue with a fine apical androconial spot; very scalloped wings; FEMALE ground color bright blue with well-defined black marginal bands; veins darkened
Subspecies 6 described

OGYRIS ZOSINE TYPHON
Purple Azure

Size	Zone	Status
2⁵⁄₈ in • 65mm	6	Not protected

Habitat & Ecology frequently encountered in open woods where *Phoradendron* (mistletoe) occurs
Other Characteristics BELOW forewing ground color dark brown with white at base, subapex, and broadly along lateral margin; three dark brown spots outlined in iridescent blue; dark purplish brown hindwing; MALE above, unmarked; moderately iridescent blue
Subspecies 4 described

Genus Palaeochrysophanus

These dimorphic butterflies with highly colored males occur in temperate Europe and Asia.

Genus Paraphnaeus

This genus of tropical African hairstreaks is closely related to the genus *Aphnaeus*.

▲

PALAEOCHRYSOPHANUS HIPPOTHOE
Purple-edged Copper

Size	Zone	Status
1³/₄ in • 40mm	3	Not protected

Habitat & Ecology flowery meadows; larval food plant: *Rumex* (dock)
Other Characteristics BELOW ground color gray brown; fine black markings outlined in white; forewing pale orange at base; hindwing darker, overscaled with blue at base; FEMALE above, paler; black hindwing; orange outer wing margin
Subspecies 15 described

Genus Parrhasius

This genus of North and Central American hairstreaks is blue above.

▲

PARRHASIUS M-ALBUM
White-M Hairstreak

Size	Zone	Status
1¹/₈ in • 30mm	1•2	Not protected

Habitat & Ecology denizen of open oak woods
Other Characteristics ABOVE ground color blue with dark forewing apices; black wing margins; FEMALE similar, with broader wing margins
Subspecies 3 described

PARAPHNAEUS HUTCHINSONII
Silverspot, Hutchinson's Highflier

Size	Zone	Status
1³/₄ in • 40mm	4	Not protected

Habitat & Ecology woodland in scrub, savannah, and montane areas; larval food plants: include *Acacia robusta* (member of the acacia family)
Other Characteristics ABOVE ground color blue with silver spots below repeated as white above; FEMALE very similar
Subspecies 2 described

Genus Phaeostrymon

This genus consists of a single species in the southwestern United States.

PHAEOSTRYMON ALCESTIS
Soapberry Hairstreak

Size	Zone	Status
1¹/₈ in • 30mm	1	Not protected

Habitat & Ecology variable, mixed oak woods and waysides; larval food plant: *Sapindus saponaria* (soapberry)
Other Characteristics ABOVE ground color uniform, unmarked brown; FEMALE similar
Subspecies 2 described so far

Genus Philiris

This is a genus of blues with numerous species endemic to the Australian region, especially New Guinea.

PHILIRIS HELENA

Size	Zone	Status
1³/₈ in • 35mm	6	Not protected

Habitat & Ecology infrequently encountered in secondary forests and disturbed habitats such as old gardens; larval food plant: *Macaranga* (Euphorbiaceae family)
Other Characteristics BELOW ground color unmarked silver gray; FEMALE darker brown with a small median patch on forewing either white or pale blue; below, similar to male
Subspecies 5 described

Genus Phasis

Endemic primarily to South Africa, these butterflies are distinguished by their markings and wing shape.

PHASIS THERO
Hooked Copper

Size	Zone	Status
2¹/₈ in • 55mm	4	Not protected

Habitat & Ecology generally coastal scrub
Other Characteristics BELOW ground color mottled dark brown, overscaled with white along margins; FEMALE orange markings larger
Subspecies 2 described

Lycaenidae

Genus Plebejus

This is a holarctic genus of small blues that are generally dimorphic. These butterflies are exceedingly common where they occur.

PLEBEJUS ARGUS
Silver-studded Blue

Size	Zone	Status
1¹/₈ in • 30mm	3	Not protected

Habitat & Ecology open grassy areas up to moderate elevations
Other Characteristics BELOW ground color yellowish gray with pale blue at base; forewing with black cell end spot and faint orange submarginal band; FEMALE above, ground color brown with sprinkling of blue basally on wings
Subspecies 38 (and many aberrations) described

PLEBEJUS SAEPIOLUS
Greenish Blue, Greenish Clover Blue

Size	Zone	Status
1¹/₈ in • 30mm	1	Not protected

Habitat & Ecology variable from waysides to grassy meadows
Other Characteristics ABOVE ground color blue with greenish overtones darker, toward wing margins; FEMALE generally dimorphic; ground color brown to gray brown
Subspecies 6 described

Genus Plebicula

This genus of blues is widely distributed in Europe to Russia and North Africa.

PLEBICULA ATLANTICA WEISSEI
Atlas Blue

Size	Zone	Status
1 in • 25mm	3	Not protected

Habitat & Ecology restricted to Morocco; dry scrub and montane rocky areas
Other Characteristics BELOW ground color yellowish gray with spot bands very prominent; with white hindwing streak; FEMALE very different; above, ground color brown with broad orange areas on both wings
Subspecies 3 described

PLEBICULA NIVESCENS
Mother of Pearl Blue

Size	Zone	Status
1³/₈ in • 35mm	3	Not protected

Habitat & Ecology flowery montane meadows
Other Characteristics BELOW ground color gray, yellowish on hindwing; sometimes blue suffusion on wing bases; postmedian spot bands distinct on both wings; cream white streak near end cell extends to margin; FEMALE above, ground color dark brown with orange yellow lunules on submargin on both fore- and hindwings
Subspecies 3 listed of doubtful validity

Genus Polyommatus

This large genus of common blues is widely distributed in Europe and Asia.

Genus Poecilmitis

These species are found in southern Africa, and are highly colored above.

POLYOMMATUS ICARUS
Common Blue

Size	Zone	Status
1³/₈ in • 35mm	3	Not protected

Habitat & Ecology variable but generally common in open sunny fields, clearings, meadows in lowlands and up to 6,000 ft (2,000m)
Other Characteristics BELOW ground color gray; bluish green basal suffusion with many spots on base of wings with a small spot in cell in base and one just below it; FEMALE above, ground color dark brown with a purplish sheen on wing bases
Subspecies 15 described

POECILMITIS THYBE
Golden Copper, Opal Copper

Size	Zone	Status
1¹/₈ in • 30mm	4	Not protected

Habitat & Ecology variable, but common in coastal dune localities and montane localities
Other Characteristics ABOVE ground color golden orange; FEMALE larger orange areas; reduced blue areas

Genus Pseudalmenus

This is a genus of Australian and Tasmanian hairstreak butterflies.

♀
▲

PSEUDALMENUS C. CHLORINDA
Tasmanian Hairstreak

Size	Zone	Status
1¹/₈ in • 30mm	6	Not protected

Habitat & Ecology open woodland; larval food plants: *Leguminosae* (legume family)
Other Characteristics BELOW ground color dark brown; green gold median band across forewing, with marking at end cell; orange tornal patch on hindwing; MALE duller; apices more acute
Subspecies 4 described

Genus Pseudaricia

The 3 species of this genus are temperate Eurasian blue butterflies that inhabit mountain meadows.

PSEUDARICIA NICIAS
Silvery Argus

Size	Zone	Status
1 in • 25mm	3	Not protected

Habitat & Ecology lowland flowery meadows in northern Europe; mountains in southern Europe up to 5,000 ft (1,500m); larval food plant: *Geranium* (geranium)
Other Characteristics BELOW ground color pale gray; indistinct submarginal markings; FEMALE above, ground color brown; light brown marginal fringes
Subspecies 5 described

Genus Pseudolycaena

These 2 species are widely distributed in Central and South America.

PSEUDOLYCAENA MARSYAS

Size	Zone	Status
2 in • 50mm	2	Not protected

Habitat & Ecology dry tropical forest clearings; scrub areas; mature secondary habitats
Other Characteristics BELOW ground color purple gray; FEMALE paler; black forewing border
Subspecies 2 described

Genus Pseudophilotes

This small genus is characterized by checkered fringes on both wings.

Genus Purlisa

These colorful Asian hairstreaks are powerful fliers.

PURLISA GIGANTEUS

Size	Zone	Status
2⁵/₈ in • 65mm	5	Rare

Habitat & Ecology montane areas
Other Characteristics BELOW ground color lustrous pale brown tan; darker on basal two-thirds of wings; tan marginal spot band on hindwing; two black eyespots and tornus sprinkled with pale blue scales; FEMALE above, extensive dark markings
Subspecies 2 described

PSEUDOPHILOTES HYLACTOR
Baton Blue

Size	Zone	Status
1 in • 25mm	3	Not protected

Habitat & Ecology flowery meadows; rocky areas; mountain slopes at 7,000 ft (2,500m); larval food plant: *Thymus serpyllum* (thyme)
Other Characteristics BELOW ground color light powdery blue with grayish black margins; FEMALE above, ground color brownish black; reduced blue suffusion at base
Subspecies 4 described

Genus Quercusia

These 2 species of hairstreak in this genus are endemic to temperate Europe and Asia.

QUERCUSIA QUERCUS
Purple Hairstreak

Size	Zone	Status
1¹/₈ in • 30mm	3	Not protected

Habitat & Ecology oak woodlands up to 5,000 ft (1,500m); larval food plant: *Quercus* (oak)
Other Characteristics BELOW ground color shiny purplish blue; black at apex; broad black borders on both wings; FEMALE purplish blue coloration more vivid; with enlarged dark broad margins; below, markings reduced
Subspecies 4 described

Genus Rapala

This showy genus from Southeast Asia are normally brightly colored above.

RAPALA IARBUS
Common Red Flash

Size	Zone	Status
1³/₄ in • 40mm	5	Not protected

Habitat & Ecology paths; edges of lowland, and upland primary and secondary forest; larval food plant: *Nephelium lappaceum* (rambutan)
Other Characteristics BELOW ground color grayish tan; hindwing overscaled with blue on tornus and a single black spot; FEMALE duller
Subspecies 4 described

Genus Rekoa

This small genus of neotropical species is found from Mexico southward but occasionally enter North America.

♀

REKOA MARIUS

Size	Zone	Status
1¹/₈ in • 30mm	1•2	Not protected

Habitat & Ecology dry tropical forests; mature secondary forests; little known about its biology
Other Characteristics ABOVE ground color dull brown; steel blue rays over hindwing; a distinct red thecla spot on the hindwing tornus; MALE above, brilliant blue; narrow apex on forewing
Subspecies none described

Genus Satyrium

With nearly 50 species, this large genus is endemic to Eurasia and North America.

♀
▲

SATYRIUM ACACIAE
Sloe Hairstrak

Size	Zone	Status
1¹/₈ in • 30mm	3	Not protected

Habitat & Ecology open woodlands and forest edges where *Prunus spinosa* (sloe) occurs
Other Characteristics ABOVE ground color dark brown with a small, tornal hair tuft on hindwing; MALE below, small orange spot near hindwing tornus; lacks androconial patch on forewing
Subspecies 5 described

SATYRIUM ACADICA
Acadian Hairstreak,
Northern Willow Hairstreak

Size | **Zone** | **Status**
1 1/8 in • 30mm | 1 | Not protected

Habitat & Ecology open woodland; marshes, moist areas; larval food plant: *Salix* (willow)
Other Characteristics ABOVE ground color uniform dark brown; with a thecla spot and sexual patch in forewing cell; FEMALE lacks sex patch
Subspecies 4 described

SATYRIUM BEHRII
Behr's Hairstreak, Orange Hairstreak

Size | **Zone** | **Status**
1 1/8 in • 30mm | 1 | Not protected

Habitat & Ecology variable, in slopes; scrubby flats; arid canyons; washes; larval food plant: *Hipparchia* (antelope brush)
Other Characteristics BELOW ground color grayish brown; well developed black bars at end of cells; orange-capped eyespot and black tornal marking on hindwing margin; FEMALE above, darker; lacks the scent patch in forewing cell
Subspecies 3 described

SATYRIUM CALANUS
Banded Hairstreak

Size | **Zone** | **Status**
1 1/8 in • 30mm | 1 | Not protected

Habitat & Ecology open habitats; oak woodlands near shale barrens; limestone ridges; larval food plant: *Quercus* (oak)
Other Characteristics ABOVE ground color unmarked brown; gray brown scent patch in forewing cell; FEMALE paler; lacks scent patch
Subspecies 4 described

Lycaenidae

SATYRIUM EDWARDSII
Edward's Hairstreak,
Scrub Oak Hairstreak

Size	**Zone**	**Status**
1 1/8 in • 30mm	1	Not protected

Habitat & Ecology scrub; open woodlands; larval food plant: *Quercus* (oak)
Other Characteristics ABOVE rich brown; large blue spot on hindwing; FEMALE similar
Subspecies none described

SATYRIUM ILICIS
Ilex Hairstreak

Size	**Zone**	**Status**
1 3/8 in • 35mm	3	Not protected

Habitat & Ecology open woodlands on rough, montane slopes to 5,000 ft (1,500m); larval food plant: *Quercus* (oak)
Other Characteristics BELOW ground color pale brown; white postmedian band broken and edged on inner margin with black; MALE above, lacks tawny overscaling; below, reduced orange submarginal band
Subspecies 6 described

♀

SATYRIUM LIPAROPS
Striped Hairstreak

Size	**Zone**	**Status**
1 3/8 in • 35mm	1	Not protected

Habitat & Ecology woods near watercourses; infrequently encountered but commonly observed at flowers; larval food plants: include *Quercus*, (oak) and *Salix* (willow)
Other Characteristics BELOW ground color gray brown; darker forewing markings etched in white and separated by thin white stripes; FEMALE lacks scent patch
Subspecies 4 described

SATYRIUM W-ALBUM
White-letter Hairstreak

Size	Zone	Status
1¹/₈ in • 30mm	3	Not protected

Habitat & Ecology brush; open woodland; isolated clumps of trees; avid flower visitor; larval food plants: include *Ulmus* (elm)
Other Characteristics ABOVE ground color uniform light brown; MALE much darker brown on both surfaces; above, scent patch in the forewing cell
Subspecies 3 described

Genus Spindasis

These African and Asian hairstreaks have distinctive color combinations of brown, orange, and blue.

Genus Scolitantides

This genus of 3 small blues is generally encountered in open habitats.

SCOLITANTIDES BAVIUS
Bavius Blue

Size	Zone	Status
1¹/₈ in • 30mm	3	Not protected

Habitat & Ecology flowery open meadows; rocky slopes up to 3,000 ft (1,000m); avid flower visitor
Other Characteristics BELOW ground color gray; with prominent black spots, especially on forewing; marginal wing fringes checkered; red orange submarginal spot band on hindwing; FEMALE above, black; blue restricted to base of wings; orange lunules on hindwing
Subspecies 7 described

SPINDASIS NATALENSIS
Natal Barred Blue

Size	Zone	Status
1¹/₈ in • 30mm	4	Not protected

Habitat & Ecology scrub and bushy habitats, savannah; larvae are associated with ants
Other Characteristics BELOW ground color pale yellow; brick red markings with metallic gold spots; 2 brick red marginal lines on forewing; gold submarginal band on hindwing extends to tornus with two black spots; FEMALE 2 or more white spots in forewing cell; above, paler with thin white marginal line above tornus
Subspecies none described

Genus Strymon

The 30 hairstreaks in this genus are normally characterized by a white abdomen in the males, and a brown abdomen in females.

STRYMON ALBATA
White Hairstreak

Size	Zone	Status
1¹/₈ in • 30mm	1•2	Not protected

Habitat & Ecology open country, large forest clearings; migratory, probably reintroduced into the USA annually; larval food plants: *Malvaceae* (mallow family)
Other Characteristics ABOVE ground color basically white; forewing costa broadly blackish brown; some black spots on the hindwing; MALE above, large black scent patch in the forewing cell
Subspecies 2 described

STRYMON ACIS
Drury's Hairstreak, Acis Hairstreak, Antillean Hairstreak

Size	Zone	Status
1 in • 25mm	1•2	Not protected

Habitat & Ecology open woods; scrub; open trails; avid flower visitor; larval food plant: *Croton* (croton)
Other Characteristics ABOVE ground color dark brown with a small orange lunule on the hindwing tornal area; FEMALE abdomen is not white but brown
Subspecies 7 described

STRYMON BEBRYCIA
Mexican Gray Hairstreak

Size	Zone	Status
1¹/₈ in • 30mm	1•2	Not protected

Habitat & Ecology secondary disturbed habitats
Other Characteristics ABOVE brownish gray; variable with a few blue to a blue patch of scales at base of wings; hindwing with dusky gray blue patch on lower half of hindwing; FEMALE similar
Subspecies 2 described

Genus Tajuria

This genus of more than 30 species is restricted to Southeast Asia. Although brightly colored above, they are dull and nondescript below.

TAJURIA MANTRA

Size	Zone	Status
1³/₈ in • 35mm	5•6	Not protected

Habitat & Ecology lowland tropical forests; gardens; larval food plant: *Loranthus* (mistletoe)
Other Characteristics BELOW ground color grayish brown; faint submarginal spot band on forewing; two black spots on hindwing; dusted with blue; blue with black margins broader; FEMALE above, dull blue; hindwing with two black spots near tornus; below, larger orange patch
Subspecies 7 described

Genus Thecla

There is debate over the classification of many of these species, but typically they have squarer wings than many of the hairstreaks.

THECLA BETULAE
Brown Hairstreak

Size	Zone	Status
1³/₈ in • 35mm	3	Not protected

Habitat & Ecology scrubby areas; open woodland; larval food plants: include *Prunus* (plum family), *Betula* (birch)
Other Characteristics ABOVE ground color dark brown; median cell end orange patch with a smaller patch of orange on the tornus near the tails of the hindwings; FEMALE lacks all orange except near the tornus and on the tails
Subspecies 4 described

THECLA ELONGATA

Size	Zone	Status
1¹/₈ in • 30mm	2	Not protected

Habitat & Ecology open clearings; paths; trails in highland subandean forests
Other Characteristics BELOW ground color dark brownish black; forewings with a small bright blue patch below cell; hindwings with blue in cell and beyond; FEMALE unknown
Subspecies none described

THECLA GABATHA

Size	Zone	Status
1⁷/₈ in • 45mm	2	Not protected

Habitat & Ecology tropical forest edges; paths
Other Characteristics BELOW ground color grayish tan; hindwing with postmedian white band edged in red; black eyespot capped in reddish orange near upper tail; a black spot on tornus; FEMALE duller in coloration
Subspecies none described

Genus Theclopsis

This is a small genus of neotropical hairstreaks that are characterized by the lustrous blue above and silver coloration on the under surface.

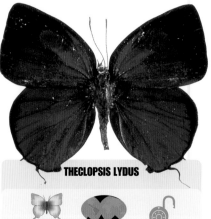

THECLOPSIS LYDUS

Size	Zone	Status
1¹/₈ in • 30mm	2	Not protected

Habitat & Ecology rain forest edges; paths; little known on its biology
Other Characteristics BELOW gray forewing with faint darker postmedian band and androconial patch near inner margin; hindwing with irregular black postmedian band and a white marginal line; a black eyespot encircled in red and a black spot on tornus; FEMALE duller blue; broader black margins; lacks the scent patch
Subspecies 2 described

Genus Theritas

These are colorful hairstreaks with unique patterns below.

THERITAS CORONATA

Size	Zone	Status
2¹/₂ in • 63mm	2	Not protected

Habitat & Ecology rain forest; populations localized
Other Characteristics BELOW black, overscaled with green; forewing with black postmedian line overlaid with lustrous gray; hindwing outer margin green at base edged in chartreuse; black and red postmedian lines; tornus black; FEMALE larger red spots along the hindwing tornus; below, chartreuse line on hindwing less vibrant
Subspecies none described

Genus Thersamolycaena

This is a genus of British coppers in which the sexes are markedly dimorphic.

THERSAMOLYCAENA DISPAR
Large Copper

Size	Zone	Status
1³/₄ in • 40mm	3	Protected

Habitat & Ecology fens; marshy areas; larval foodplant: *Rumex* (dock)
Other Characteristics BELOW gray tan along forewing margin; hindwing ground color gray, suffused with blue on wing bases, MALE above, virtually unmarked; narrow dark margins on both wings; a large cell spot on forewing
Subspecies none described

Genus Thersamonia

Related to the *Lycaena*, these Eurasian coppers are very sexually dimorphic.

Genus Tmolus

These marked dimorphic lycaenids are characterized by the brick red bands and spots on the undersurface.

THERSAMONIA THERSAMON
Lesser Fiery Copper

Size	Zone	Status
1¹/₈ in • 30mm	3	Not protected

Habitat & Ecology moist areas among scrub; light or open woodlands up to 4,000 ft (1,200m); larval food plant: *Rumex* (dock)
Other Characteristics ABOVE forewing yellow, with orange and gray with gray near margin; MALE below, uniform brilliant copper; narrow black border on forewing; hindwing suffused with grey purple
Subspecies 3 described

TMOLUS ECHION
Large Lantana, Four-spotted Hairstreak

Size	Zone	Status
1 in • 25mm	1•2•6	Not protected

Habitat & Ecology disturbed habitats; larval food plant: *Lantana* (shrub verbena); was introduced into Mexico and Hawaii to control this opportunistic plant
Other Characteristics ABOVE ground color black with bright blackish blue suffusion on hindwings; FEMALE above, dark brown
Subspecies 2 described

Genus Vacciniina

This genus of blues is restricted to and around the Arctic Circle. They feed on *Vaccinium* (blueberry) as larvae.

VACCINIINA OPTILETE
Yukon Blue, Cranberry Blue

Size	Zone	Status
$7/8$ in • 20mm	1•3	Not protected

Habitat & Ecology moorland; moist mountain meadows where *Vaccinium* (blueberry) occurs
Other Characteristics ABOVE ground color brilliant violet blue; FEMALE above, dull violet blue ground color with broad dark margins
Subspecies 12 described

Genus Vaga

The 2 species of this genus are noted for their remarkable coloration.

♀

VAGA BLACKBURNI
Green Hawaiian Blue, Blackburn's Blue

Size	Zone	Status
1 in • 25mm	6	Not protected

Habitat & Ecology Hawaiian forests; avid flower visitor; lowland areas; larval food plant: *Acacia* (mimosa tree)
Other Characteristics ABOVE ground color blackish brown, with blue overscaling at wing bases; MALE above, ground color deep bluish purple
Subspecies none described

Genus Xamia

These Central American hairstreaks are characterized by tails on the hindwing margin.

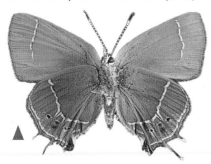

XAMIA XAMI
Xamia, Succulent Hairstreak

Size	**Zone**	**Status**
1 in • 25mm	1•2	Not protected

Habitat & Ecology sunny xeric montane areas; canyon sides in coniferous and oak-pine forests
Other Characteristics ABOVE lustrous golden tan; black borders; scent patch in forewing cell; veins darkened near outer margin of hindwing; FEMALE veins more evident; larger black margins
Subspecies 2 described

Genus Zizina

This genus of species comprises rather unusual blues that are restricted to Africa, Asia, and Australia.

Genus Zizula

These very small blue butterflies are restricted to North and South America.

♀

ZIZINA OTIS
Common Grass Blue, Lucerne Blue, Clover Blue, Bean Blue, Lesser Grass Blue

Size	**Zone**	**Status**
1 in • 25mm	5•6	Not protected

Habitat & Ecology open grassy areas with flowers; larval food plant, *Mimosa pudica* (sensitive plant)
Other Characteristics BELOW darker gray tan end cell spot; FEMALE above, purplish blue restricted to basal two thirds and black margins expanded
Subspecies 12 described

ZIZULA CYNA
Cyna Blue

Size	**Zone**	**Status**
1 in • 25mm	1•2	Not protected

Habitat & Ecology mesic habitats; desert scrub; they move about but are not truly migratory; larval food plant: flower buds of *Acanthaceae* (acanthia family)
Other Characteristics BELOW ground color tan with black postmedian and submarginal spot bands on both wings; MALE above, bright blue with narrower black borders
Subspecies none described

CHAPTER 5

Riodinidae

These are the metalmark butterflies, so called because of the remarkable number of iridescent patterns on both wing surfaces. Often these butterflies have been included as part of the Lycaenidae or the superfamily Lycaenoidea. However, sufficient morphological and life history differences are present in the Riodinidae to warrant family status. For example, in male riodinids, the forelegs are reduced, while they are fully functional walking legs in the Lycaenidae. Male androconia are a rare feature in riodinids but are often present in the lycaenids. There are many other features, including structural differences in the caterpillars and pupae, that are unique to riodinids. Another characteristic of most riodinids is behavioral, with most species sitting or "perching" under leaves, often with the wings outstretched.

Genus Abisara

Species within this genus usually have brown wings with purple or crimson suffusion, and the forewing apex is quite square.

ABISARA SAVITRI
Malay Tailed Judy

Size	**Zone**	**Status**
2¹/₈ in • 55mm	5	Not protected

Habitat & Ecology shade dwellers in rain forest understorey; larval food plant: probably *Mysinaceae* (river mangrove)
Other Characteristics BELOW paler rust areas overscaled with buff on proximal half; buff transverse band and forewing subapical patch; FEMALE above, white forewing transverse band extends from costa to inner margin
Subspecies 8 described

Genus Adelotypa

Usually found in the Guianas and coastal Brazil.

ADELOTYPA PENTHEA

Size	**Zone**	**Status**
1³/₈ in • 35mm	2	Not protected

Habitat & Ecology tropical deciduous and rain forest; often observed perching on tree trunks
Other Characteristics BELOW markings similar, but overscaled with white, especially on lower half of forewing; FEMALE dimorphic; ground color white, darker at forewing apex and along lateral margin
Subspecies 2 described

Genus Alesa

Endemic to the Amazon basin and the Guianas, this genus is markedly sexually dimorphic with few species represented in eastern South America.

♀

ALESA AMESIS
Green Dragon

Size	**Zone**	**Status**
2 in • 50mm	2	Not protected

Habitat & Ecology paths and clearings of rain forest understory in Amazon basin and Guianas
Other Characteristics BELOW markings similar to above, but paler; MALE dimorphic; above, blue black wings; below, similar to female
Subspecies 2 described

Genus Ancyluris

These are some of the most distinctive metalmarks, with the striking color combinations on both wing surfaces.

ANCYLURIS FORMOSISSIMA VENERABILIS

Size	Zone	Status
1³/₄ in • 40mm	2	Not protected

Habitat & Ecology rain forest
Other Characteristics ABOVE broader white median bands complete on both wings; FEMALE lacks iridescent blue and white; median bands broader on both wings
Subspecies 2 described

ANCYLURIS JURGENSENII

Size	Zone	Status
2 in • 50mm	2	Not protected

Habitat & Ecology paths and trails of primary rain forest, or mature secondary forests
Other Characteristics BELOW ground color black with blue median band; reduced submarginal transverse bands; red blotch mid-hindwing inner margin; FEMALE above, ground color brown with broad white median bands on both wings
Subspecies 2 described

Genus Apodemia

One of the most frequently encountered metalmarks in the field, this group is widely distributed in the western United States and Central America to Costa Rica.

APODEMIA MORMO
Mormon Metalmark

Size	Zone	Status
1¹/₈ in • 30mm	1	Endangered

Habitat & Ecology deserts; dunes; larval food plant: *Polygonum* (bistorts)
Other Characteristics BELOW overscaled with white on forewing apex and margins; FEMALE similar
Subspecies more than 5 described

APODEMIA PALMERI
Gray Metalmark, Mesquite Metalmark

Size	Zone	Status
7/8 in • 20mm	1•2	Not protected

Habitat & Ecology observed on plains and dry washes; larval food plant: *Prosopis juliflora* (mesquite)
Other Characteristics BELOW ground color pale gray brown with semi-transparent markings enlarged; FEMALE overall coloration duller; white marginal markings enlarged
Subspecies none described

Genus Argyrogrammana

Comprised of more than 25 species, these butterflies are small and rather delicate. Speckled markings below often combine with yellow or orange and turquoise blue.

Genus Astraeodes

This genus is widely distributed throughout the Amazon basin south to Bolivia and Peru.

ARGYROGRAMMANA ALSTONII

Size	Zone	Status
1 1/8 in • 30mm	2	Not protected

Habitat & Ecology variable, from rain forest canopy at moderate altitudes, to ridge-top trails in cloud forests up to 4,500 ft (1,500m)
Other Characteristics BELOW ground color grayish white with small black and dull iridescent aquamarine bars on forewing and anterior hindwing; gray along inner margin of forewing; FEMALE unknown
Subspecies none described

ASTRAEODES AREUTA

Size	Zone	Status
1 3/4 in • 40mm	2	Not protected

Habitat & Ecology variable, but infrequently encountered in tropical deciduous forests and rain forest
Other Characteristics BELOW similar but with black marginal spots near hindwing apex and at anal angle; FEMALE similar
Subspecies none described

Genus Calephelis

This genus has rust brown and characteristic silver blue metallic flecks, and submarginal and marginal spot bands above. These butterflies are difficult to identify.

CALEPHELIS NEMESIS
Mexican Metalmark, Fatal Metalmark

Size	Zone	Status
1 in • 25mm	1•2	Not protected

Habitat & Ecology often observed in dry washes and gullies, especially during the heat of the day; little known about life history and ecology

Other Characteristics ABOVE markings similar but ground color various shades of warm brown with darker median band on both wings;

FEMALE markings similar; above variable, ground color brown to orange brown, median area can be darker; metallic bands are subdued

Subspecies 4 described

Genus Calydna

With more than 25 species described, this genus of sexually dimorphic metalmarks are widespread throughout Central and South America.

CALYDNA CALAMISA

Size	Zone	Status
1³/₄ in • 40mm	2	Not protected

Habitat & Ecology generally observed along paths and trails in lowland forests or may visit moist sand along streams

Other Characteristics BELOW brown with white markings; hindwing with black marginal spot at apex and anal angle; FEMALE dimorphic; markings similar to male but ground color light, warm brown, with darker brown postmedian patches

Subspecies none described

Genus Cartea

These butterflies have remarkably vivid color combinations and are endemic to the western Amazon basin.

CARTEA VITULA TAPAJONA

Size	Zone	Status
1³/₄ in • 40mm	2	Not protected

Habitat & Ecology tropical rain forest of the Amazon basin and Peru
Other Characteristics BELOW black areas replaced by blackish brown; FEMALE similar but black areas replaced by brown, and red replaced by dull orange
Subspecies 3 described

Genus Chorinea

With more than 7 species described, these butterflies have palpi that are markedly reduced into a cone.

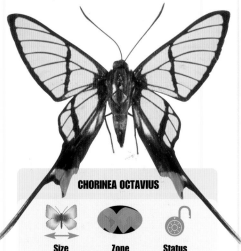

CHORINEA OCTAVIUS

Size	Zone	Status
1³/₄ in • 40mm	2	Not protected

Habitat & Ecology paths, rain forest clearings; often observed under leaves of herbaceous plants; larval food plant: *Prionostemma aspera* (member of the hippocratea family)
Other Characteristics BELOW similar; FEMALE larger; red patch on hindwing may be enlarged
Subspecies 3 described

Genus Dodona

Composed of 15 species, this Asian genus is endemic to the Himalayas and montane areas of northeastern India and Burma.

DODONA DIPOEA
Lesser Punch

Size	Zone	Status
1⁷/₈ in • 45mm	5	Not protected

Habitat & Ecology infrequently encountered in montane forests and areas between 4,000 and 8,000 ft (1,200 and 2,400m)
Other Characteristics BELOW paler ground color; FEMALE forewing apices rounded
Subspecies 2 described

Size	Zone	Status
1³/₄ in • 40mm	2	Not protected

Habitat & Ecology along paths and open clearings in primary rain forests

Other Characteristics BELOW paler; FEMALE duller brown ground color; broad orange forewing apical bar; horizontal orange patch in basal cell to anal vein; orange discal patch on hindwing with orange/white submarginal spot band

Subspecies 4 described

Genus Esthemopsis

Wing patterns of this neotropical species can vary from tiger stripes to arctiid moth mimics.

Genus Eurybia

With more than 25 species recognized in this neotropical genus, it is one of the largest groups in the Riodinidae.

EURYBIA JUTURNA HARI

Size	Zone	Status
2⁵/₈ in • 65mm	2	Not protected

Habitat & Ecology trails and paths in tropical deciduous and rain forest of Peru and Bolivia

Other Characteristics ABOVE markings and coloration similar but darker without forewing eyespot and orange hindwing patch more vivid; FEMALE above, eyespot more evident and postmedian and submarginal spot bands complete on both wings

Subspecies 3 described

EURYBIA LYCISCA

Size	Zone	Status
2³/₈ in • 60mm	2	Not protected

Habitat & Ecology primary and secondary rain forest up to 1,500 ft (500m); males are very territorial; larval food plants: include various *Marantaceae* (marantia family)

Other Characteristics BELOW paler, iridescent blue on hindwing; FEMALE above, faint indication of blue

Subspecies none described

Genus Euselasia

There are probably more than 100 species in this neotropical genus. They are characterized generally by the overall diversity of wing patterns below.

 EUSELASIA EURITEUS

Size	Zone	Status
1³/₈ in • 35mm	2	Not protected

Habitat & Ecology rain forest clearings and trails
Other Characteristics ABOVE ground color black with deep blue forewing transverse band from mid-costa to inner margin; hindwing with blue along lateral margin; FEMALE below, similar to male; above, ground color warm brown with golden orange along anal angle of hindwing
Subspecies 2 described

Genus Hades

The coloration, flight behavior, and name suggest the sinister nature of this neotropical genus.

HADES NOCTULA

Size	Zone	Status
2¹/₈ in • 55mm	2	Not protected

Habitat & Ecology primary and secondary rain forest; avid flower visitors; larval food plant: *Spondias mombin* (yellow mombin)
Other Characteristics ABOVE ground color black, lighter on wing margins with indication of white rays on hindwing lateral margin; FEMALE similar
Subspecies none described

Genus Hamearis

Despite its common name, this genus is not a fritillary and is the only metalmark found in Europe.

HAMEARIS LUCINA
Duke of Burgundy Fritillary

Size	Zone	Status
1³/₈ in • 35mm	3	Not protected

Habitat & Ecology woods up to 4,000 ft (1,200m); larval food plants: *Lachenalia* (cowslip), *Primula* (primrose)
Other Characteristics BELOW ground color brown with orange at base on forewing; markings similar on forewing but with a buff subapical band and with orange submarginal band etched with buff on proximal margin; FEMALE similar
Subspecies 2 described

Genus Helicopis

Members of this colorful genus of South American metalmarks are characterized by their extraordinary tails.

HELICOPIS GNIDUS

Size	Zone	Status
2 in • 50mm	2	Not protected

Habitat & Ecology rain forest
Other Characteristics BELOW base of wings rust orange; FEMALE medial forewing patch enlarged
Subspecies none described

Genus Ithomiola

These butterflies resemble a number of ithomiid butterflies (especially *Heterosais aureola* and *Napeogenes corea*) and are probably involved in mimicry complexes.

ITHOMIOLA CASCELLA

Size	Zone	Status
1¹/₈ in • 30mm	2	Not protected

Habitat & Ecology paths in the understorey of rain forests of Colombia; infrequently encountered; populations are localized
Other Characteristics wing patterns are variable in both sexes; BELOW orange overscaling along lateral margin more extensive; MALE with orange hindwing submarginal spot band
Subspecies 2 described

Genus Lyropteryx

Divided into 5 species, members of this unusual neotropical genus are strongly sexually dimorphic.

LYROPTERYX APOLLONIA

Size	Zone	Status
2 in • 50mm	2	Not protected

Habitat & Ecology paths in lowland rain forest
Other Characteristics ABOVE red marks reduced to single spots on basal hindwing; ground color at base black with emerald green; FEMALE above, broad red hindwing marginal band; darker veins
Subspecies 4 described

Genus Mesene

The 30 or more species that make up this genus are small, but bright, colorful butterflies.

MESENE PHAREUS RUBELLA
Cramer's Mesene

Size	Zone	Status
1¹/₈ in • 30mm	2	Not protected

Habitat & Ecology lowland rain forest; observed well hidden under leaves of bushes along trails and paths; when disturbed will fly out and drop down into the bushes or on roads; larval food plant: *Paullinia pinnata* (bread and cheese)
Other Characteristics ABOVE similar, but ground color an even carmine red; FEMALE forewings slightly broader and somewhat duller or darker
Subspecies 8 described

Genus Mesosemia
With the prominent eyespot on the upper forewing, some species within this neotropical genus may appear similar to *Eurybia*. A number of species are sexually dimorphic, and there is much individual variation.

MESOSEMIA HEDWIGIS

Size	Zone	Status
1⁷/₈ in • 45mm	2	Not protected

Habitat & Ecology lowland tropical forest; larval food plant: *Sapindaceae* (soapberry family)
Other Characteristics BELOW ground color and markings paler with an eyespot in median band; enlarged eyespot in cell; single eyespot on median hindwing band; FEMALE similar
Subspecies none described

MESOSEMIA LORUHAMA

Size	Zone	Status
1⁷/₈ in • 45mm	2	Not protected

Habitat & Ecology lowland tropical deciduous forests and rain forest in Ecuador and Peru
Other Characteristics BELOW ground color brown; single eyespot in pale forewing median band; FEMALE broad white bands from forewing mid-costa to inner margin, and apex hindwing to inner margin
Subspecies 2 described

Genus Methone

These South American metalmarks are characterized by their scalloped hindwings.

METHONE CECILIA

Size	Zone	Status
1³/₄ in • 40mm	2	Not protected

Habitat & Ecology denizens of the understory in moist, cloud forests at 1,200–4,800 ft (400–1,450m); little known about its life history
Other Characteristics BELOW paler, basal two-thirds of hindwing cell darker orange; with white submarginal spot band; FEMALE orange areas yellower; forewing with curved white bar at apex; above, white spots along hindwing submargin
Subspecies 2 described

Genus Necyria

This genus is composed of more than 7 species, most of which are quite variable.

NECYRIA BELLONA

Size	Zone	Status
2 in • 50mm	2	Not protected

Habitat & Ecology rain forest; tropical deciduous forests in southern Brazil and eastern Peru
Other Characteristics ABOVE ground color blackish brown with red angular bars on wings; iridescent blue rays from margin of red bar to wing margin on both wings; FEMALE unknown
Subspecies none described

Genus Pandemos

These unusual dimorphic metalmarks are infrequently encountered in Central and South America and are favorites of collectors.

PANDEMOS PASIPHAE

Size	Zone	Status
2¹/₈ in • 55mm	2	Not protected

Habitat & Ecology lowland rain forest
Other Characteristics BELOW forewing tan and shading to lilac; FEMALE white with brown at apex
Subspecies none described

▲
♀

Genus Paralaxita

Members of this genus are strongly sexually dimorphic with a number of secondary sexual characters in the males.

Size	**Zone**	**Status**
1³/₄ in • 40mm	5	Not protected

Habitat & Ecology dense forests with bright sun; montane areas up to 4,000 ft (1,200m)
Other Characteristics ABOVE ground color brown with cerise at apex; MALE strongly dimorphic; above, ground color blue; intense color below
Subspecies 3 described

Genus Periplacis

Members of this neotropical genus are closely related to *Menander*, and are sexually dimorphic.

Genus Praetaxila

These metalmarks are found from New Guinea to the Australian mainland. They are markedly sexually dimorphic.

Size	**Zone**	**Status**
1³/₄ in • 40mm	2	Not protected

Habitat & Ecology paths and trails in rain forest; tropical deciduous forests to 1,200 ft (400m)
Other Characteristics BELOW ground color brown, darker at wing apices and adjacent submargin; FEMALE above, forewing overscaled faintly with blue; below, overscaled with tan and markings similar to above on both wings; hindwings with black marginal spot at apex, and near anal angle
Subspecies none described

  ,

Size	**Zone**	**Status**
2⁷/₈ in • 70mm	5	Not protected

Habitat & Ecology open primary and secondary forests
Other Characteristics BELOW mottled brown; forewing band cream; FEMALE strongly dimorphic
Subspecies none described

Genus Rhetus

The 4 species in in this South American genus are characterized by their long hindwings.

RHETUS ARCIUS

Size	**Zone**	**Status**
1³/₄ in • 40mm	2	Not protected

Habitat & Ecology variable, rain forest and mature secondary forests to 4,200 ft (1,300m)
Other Characteristics BELOW lacks blue; FEMALE lacks blue; larger median white band
Subspecies 7 described

Genus Semomesia

With the large forewing eyespot, this neotropical genus resembles *Mesosemia* and was originally incorporated with it.

Genus Rodinia

This South Ameican genus has irregular-shaped hindwings.

RODINIA BARBOURI

Size	**Zone**	**Status**
2¹/₈ in • 55mm	2	Not protected

Habitat & Ecology sunny patches on low rain forest trails; can be observed hilltopping
Other Characteristics BELOW pale brown at apex and along wing margins, shading to dark brown on posterior half of hindwing; FEMALE similar
Subspecies none described

SEMOMESIA CAPANEA

Size	**Zone**	**Status**
1¹/₈ in • 30mm	2	Not protected

Habitat & Ecology coastal forests in Guiana and rain forests in Peru and central Brazil
Other Characteristics BELOW lacks eyespot, blue extends to wing margins; FEMALE dimorphic; ground color brown; forewing with white band from mid-costal to inner margins; hindwing with concentric pattern of male present, but dark brown and tan overscaled with white
Subspecies 4 described

Riodinidae

Genus Stalachtis
These tiger metalmarks possess a stout thorax and abdomen and mimic certain arctiid moths and danaids with their slow, fluttering flight. Polymorphism is present in both sexes of this genus.

STALACHTIS CALLIOPE

Size
2⅝ in • 65mm

Zone
2

Status
Not protected

Habitat & Ecology paths and open trails in tropical moist and dry rain forest
Other Characteristics BELOW similar; FEMALE can be variable, lacks hindwing postdiscal orange spot band; may have more yellow on the forewing subapical area
Subspecies 3 described

STALACHTIS PHLEGIA

Size
2 in • 50mm

Zone
2

Status
Not protected

Habitat & Ecology tropical forest edges and clearings; common in Guiana and Brazil
Other Characteristics BELOW similar; with forewing rust submarginal band and thinner hindwing submarginal band; FEMALE forewing apices rounded; below, submarginal rust bands more intense
Subspecies 3 described

Genus Stiboges
This genus with its single species is widely distributed from north India and west China to Malaysia and parts of Indonesia.

STIBOGES NYMPHIDIA

Size
1¾ in • 40mm

Zone
5

Status
Not protected

Habitat & Ecology dense forests at moderate elevations with populations localized
Other Characteristics BELOW white submarginal and marginal spot bands more extensive
FEMALE similar; forewing apex more rounded
Subspecies 4 described

Genus Synargis

This genus is sexually dimorphic and often confused with *Juditha* and *Nymphidium*.

♀

SYNARGIS NYCTEUS

Size	Zone	Status
2 in • 50mm	2	Not protected

Habitat rain forest at sea level up to 2,400 ft (800m)
Other Characteristics BELOW ground color, blackish brown on apices and along lateral margin on both wings; MALE forewing entirely rust brown with similar barred markings of female
Subspecies none described

Genus Syrmatia

These small, unique metalmarks with their elongated tails resemble some of the larger wasps in flight.

SYRMATIA DORILAS

Size	Zone	Status
1¹/₈ in • 30mm	2	Not protected

Habitat & Ecology rain forest slopes, generally in association with streams; populations are somewhat localized
Other Characteristics BELOW paler; FEMALE similar with the forewing red and white bars enlarged; hindwing has median white bar
Subspecies none described

Genus Thisbe

These neotropical butterflies are distinguished by the broad white bands and patches and the characteristic hindwing shape.

THISBE LYCORIAS

Size	Zone	Status
1³/₄ in • 40mm	2	Not protected

Habitat & Ecology edges; paths; rain forest clearings; secondary forests up to 3,600 ft (1,200m); larval food plant: *Cassia alata* (emperor's candlesticks)
Other Characteristics BELOW paler with bars at forewing apex and along submargin of wings enlarged; FEMALE white areas larger
Subspecies 3 described

glossary

Italics refer to cross referenced terms within the glossary.

adult The reproductive stage of the butterfly, with a proboscis to take food, three pairs of legs, four wings, and genitalia are present.

anal angle The *posterior* portion of the lateral *margin* of the wing; often equals the *tornus*.

androdoconia Specialized scales that produce chemicals or *pheromones* on various areas of the wings; these are used in courtship to attract females.

anterior Before, or pertaining to the front part of the body or a structure, such as the wing.

apex The outer point of the wing along the *costal margin*.

austral Southern, or of the south.

Australian Geographic area that includes Australia, New Zealand, New Guinea and eastward into the Pacific.

basal Proximal, or area of the wing closest to the body.

Batesian mimicry A complex of species that are similar in appearance, some of which may be afforded protection through chemical defense against predators.

boreal Northern, or of the north.

calcareoous Growing on limestone.

canopy The top storey, or area of any forest.

cell The central portion of the wing from which most of the veins arise; it is generally closed.

claspers Valvae; a part of the male genitalia that holds the female abdomen during copulation.

cloud forest A *montane* or *submontane* forest in the tropics that is usually enveloped and derives most of its moisture from fog.

costa The *anterior margin* of the butterfly wing.

crepuscular Flying at dawn or dusk.

disc Middle portion of the wing that includes the *cell*. Also synonym for *medial*.

dimorphism Essentially defined as two forms but it generally refers to the different wing coloration and patterns between the sexes.

disturbed habitats These are habitats that have been altered from the original state; generally the original plants (flora) and animals that inhabit these areas are not as specialized as those in virgin or primary habitats.

extradiscal Area of the wing outside the *cell*.

gregarious Butterflies that congregate together at mud puddles or at *larval* food plants.

holarctic The temperate and arctic zones of Eurasia and North America.

hilltopping Butterflies that fly around or over the tops of mountains; generally these are flying along set territories and looking for females.

larva The caterpillar, and stage between the egg and the pupa, characteristically with four or more instars; also the stage of the butterfly that is that is characterized by the ability to eat solid food.

local/localized Butterfly colonies that occur in very restricted or limited habitats as opposed to some species that range widely for several miles.

lunule Crescent-shaped marking.

margin The outer edge of the wing; may refer to the *costal*, lateral, or inner wing *margins*.

medial Middle portion of the wing that includes the *cell*. Also a synonym for *disc*.

mesic Refers to a moist habitat as opposed to one that is drenched such as a rain forest.

mimicry One species that resembles another, that in turn possesses some form of chemical protection from predation.

mimicry complex Whole suite or series of species that have similar wing coloration and/or patterns.

Mullerian mimicry Species with similar wing coloration and pattern that all possess some form of chemical defense that provides protection from predators.

monotypic A genus comprised of a single species.

montane Of, or associated with mountains or mountain slopes.

nearctic The New World temperate and arctic zone comprising North America, north of Mexico.

neotropical Refers to species associated with habitats in Central and South America.

oriental Geographic region that includes southern Asia to just west of Sulawesi.

overscaled A second layer of scales on the wings that may produce subtle changes in wing coloration or enhance the color depth.

palearctic The northern temperate parts of the Old World, including Europe, most of northern Asia, and North Africa.

paramo Usually cooler, upper *montane* grasslands in Central and South America with or without limited trees, but not above timber line.

patrolling Males flying back and forth along a designated trail or areas in search of mates.

pheromone a chemical substance produced by an animal to stimulate other individuals of the same species.

polymorphic A species with multiple wing coloration patterns.

posterior Behind, or pertaining to the back of the body or structure.

postmedian Outside the *medial* portion of the wing.

primary rainforest Undisturbed very wet forest with rainfall of 275 in. (7,000 mm) per year and no distinct dry season.

pupa The chrysalis or the life stage, following the caterpillar or larva stage. During this stage the tissues are completely reorganized and eventually the butterfly emerges as the adult.

Red Data Book A publication on various butterfly groups and other organisms that examines the current conservation status and provides updated information on those that require protection or are of major concern.

sagebrush Type of shrub that characterizes vast alkaline plains in the western United States.

savannah A flat plain with rather long tropical grasslands and scattered trees; most frequently found in east Africa.

secondary rainforest Rain forest that has been partially altered either through cutting, or the introduction of species characteristic of the area.

sphragis A structure secreted by the male *Parnassius* around the female genital opening during copulation which prevents subsequent matings.

subapex Below the *apex* of the wing.

subdesert Area with less than 10 in. (240 mm) of rainfall per year.

submargin The portion of the wing just inside the *margin*.

submontane Under, or at the base of a mountain or mountain range.

taiga Moist, subarctic forest characterized by the presence of conifers.

tailed morph or **tailed** A butterfly with a projection on the *tornus* or *anal angle* of the hindwing.

tamarack Moist wooded areas, characterized by the presence of larches.

thecla spot A spot that appears at the base of the tail of the hindwing, usually with a cap.

tornus The corner of the wing where the outer angle meets the inner angle.

tundra A habitat with permafrost and without trees that is usually restricted to the north arctic zone.

understorey The layer of forest vegetation between the *canopy* and the ground.

xeric May refer to a dry, forest, or desert habitat.

index

key to symbols

Fold out this flap to find an at-a-glance key explaining all the symbols used throughout the book.

KEY TO SYMBOLS

 Zone 1   **Zone 2**

 Zone 3   **Zone 4**

 Zone 5   **Zone 6**

ZONE OF ORIGIN
The map symbol indicates the butterfly's zone of origin (see page 10 for world map).

THE REGIONS ARE AS FOLLOWS:
ZONE 1, the Nearctic region; ZONE 2, the Neotropical region; ZONE 3, the Palearctic region; ZONE 4, the African region, ZONE 5 Oriental region, and ZONE 6, the Australian region

FAMILY INDICATOR
In the field butterflies can be readily sorted into the five major families. Inspection of their wing shape, venation, and number of legs gives some vital clues. Throughout this book each individual entry has a stylized shape symbol to denote family membership.

 Papilionidae   **Pieridae**   **Nymphalidae**

 Lycaenidae   **Riodinidae**

 Female specimen

Ventral view

PROTECTION STATUS
All protected butterflies are given a closed padlock symbol, and If relevant a CITES—Convention of International Trade in Endangered Species of Wild Fauna—classification. The open padlock indicates that they are unprotected. Rare or endangered species (regardless of legal status) may also be indicated.

 Not protected

 Protected

TYPE OF SPECIMEN
Most specimens show the dorsal (above) surface. The triangular symbol indicates that the opposite—the ventral (below) surface—is shown. Similarly most specimens here are male; where the female is shown, a female symbol appears in the frame.